Lecture Notes in Electrical Engineering

Volume 761

The book series *Lecture Notes in Electrical Engineering* (LNEE) publishes the latest developments in Electrical Engineering - quickly, informally and in high quality. While original research reported in proceedings and monographs has traditionally formed the core of LNEE, we also encourage authors to submit books devoted to supporting student education and professional training in the various fields and applications areas of electrical engineering. The series cover classical and emerging topics concerning:

- Communication Engineering, Information Theory and Networks
- Electronics Engineering and Microelectronics
- Signal, Image and Speech Processing
- Wireless and Mobile Communication
- Circuits and Systems
- Energy Systems, Power Electronics and Electrical Machines
- Electro-optical Engineering
- Instrumentation Engineering
- Avionics Engineering
- Control Systems
- Internet-of-Things and Cybersecurity
- Biomedical Devices, MEMS and NEMS

For general information about this book series, comments or suggestions, please contact leontina.dicecco@springer.com.

To submit a proposal or request further information, please contact the Publishing Editor in your country:

China

Jasmine Dou, Editor (jasmine.dou@springer.com)

India, Japan, Rest of Asia

Swati Meherishi, Editorial Director (Swati.Meherishi@springer.com)

Southeast Asia, Australia, New Zealand

Ramesh Nath Premnath, Editor (ramesh.premnath@springernature.com)

USA, Canada:

Michael Luby, Senior Editor (michael.luby@springer.com)

All other Countries:

Leontina Di Cecco, Senior Editor (leontina.dicecco@springer.com)

**** This series is indexed by EI Compendex and Scopus databases. ****

More information about this series at http://www.springer.com/series/7818

Xi Shao · Kun Qian · Li Zhou ·
Xin Wang · Ziping Zhao
Editors

Proceedings of the 8th Conference on Sound and Music Technology

Selected Papers from CSMT

Springer

Editors
Xi Shao
Nanjing University of Posts
and Telecommunications
Nanjing, Jiangsu, China

Li Zhou
China University of Geosciences
Wuhan, Hubei, China

Ziping Zhao
Tianjin Normal University
Tianjin, China

Kun Qian
The University of Tokyo
Bunkyo-ku, Tokyo, Japan

Xin Wang
Communication University of China
Beijing, China

ISSN 1876-1100 ISSN 1876-1119 (electronic)
Lecture Notes in Electrical Engineering
ISBN 978-981-16-1648-8 ISBN 978-981-16-1649-5 (eBook)
https://doi.org/10.1007/978-981-16-1649-5

This Springer imprint is published by the registered company Springer Nature Singapore Pte Ltd.
The registered company address is: 152 Beach Road, #21-01/04 Gateway East, Singapore 189721, Singapore

Preface

The Conference on Sound and Music Technology (CSMT)—the leading Chinese computer audition conference today—is an annual milestone event increasingly noted outside its home country. This comes, as this year's CSMT 2020 has been the third edition featuring also proceedings in English—a wise decision rendering the otherwise potentially hidden included gems more accessible to the international scene. At the same time, English co-proceedings of the series' recent three editions opened up the door for submissions for a non-Chinese speaking audience. This helped to further raise the significance of this marvellous event which is celebrating its overall 8th edition this year—a lucky number in China. And indeed, looking at the exciting and inspiring contributions ahead in this book, it seems clear that CSMT was marked by hardest work of a most outstanding organising committee and perhaps a dash of luck on the side.

As the child of the original China Sound and Music Computing Workshop (CSMCW), CSMT has indeed matured to an increasingly international event by now. A great further initiative are CSMT's quite competitive challenge papers: This year, only three out of eight challenge papers were accepted (37.5% acceptance rate). And likewise, also CSMT's overall standards are on competitive levels: This year saw overall 17 out of 37 English submissions accepted (45.9% acceptance rate) and 18 out of 33 Chinese submissions (54.5%), hence featuring an overall acceptance rate of 50.0% similar to its "big" international relatives in the field such as the IEEE ICASSP annual conference.

Beyond the impressive success in broadening up internationally, it is with particular joy to note the success in establishing the event also as a truly interdisciplinary event: besides the technical contributions and attendees, CSMT has been increasingly given attention also by the community of music artists—from only one professional musician in 2013 to more than 30 participating in this year.

The 2020 event took place on 5–8 November 2020 at North University of China, Taiyuan City, in the splendid Shanxi Province. As the previous editions, it was catering for an utmost important area in the field of computer audition at the intersection of music processing and sound processing and beyond.

Machines have long been taught to *"see"*, and they learnt to recognise and increasingly understand our spoken and written language, but clearly, a lot of work still lies ahead of us to make them *"hear"* and make sense of what their hearing in much richer ways than today. In order to fully unleash the huge potential true computer audition bears in making our everyday lives a better experience, highly interdisciplinary and international collaborations will be needed uniting expertise beyond computer science from the fields of sound and music, psychology, and engineering. CSMT is perfectly well equipped to host the best and latest findings in this respect—from China and the world and from all involved disciplines.

It is with greatest excitement that one can continue reading in the oncoming contributions—and it is with greatest excitement that one can look forward to oncoming editions of this special and significant series.

Björn W. Schuller

Organising Committee

General Chair

Xingquan Shen	North China University, China

General Co-chairs

Wei Li	Fudan University, China
Qiangbin Chen	Shanghai Conservatory of Music, China
Haifeng Li	Harbin Institute of Technology, China
Yidan Zhu	The Acoustical Society of Beijing, China
Lei Wang	Shanghai Artificial Intelligence Academic Society, China
Hong Lu	Fudan University, Shanghai Computer Society, China

Consultant Committee

Lianhong Cai	Tsinghua University, China
Ye Wang	National University of Singapore, Singapore
Dong Yu	Tencent AI Lab (Seattle), USA
Baoqiang Han	China Conservatory of Music, China
Jinqin Tian	Taiyuan Electron Musical Instrument Research Institute, China

Academic Committee Co-chairs

Xi Shao	Nanjing Posts and Telecommunication University, China
Li Zhou	Chinese Geology University (Wuhan), China

LNEE Proceeding Editors

Xi Shao	Nanjing Posts and Telecommunication University, China
Kun Qian	The University of Tokyo, Japan
Li Zhou	Chinese Geology University, Wuhan, China
Xin Wang	Communication University of China, China
Ziping Zhao	Tianjin Normal University, China

Technical Programme Committee Co-chairs

Kejun Zhang	Zhejiang University, China
Mengyao Zhu	Huawei Technology, China
Zhongzhe Xiao	Suzhou University, China

Website and System Administrators

Feng Li	Shanghai Science and Technology Literature Press, China
Ke Chen	UCSD, USA
Kaikai Li	North China University, China
Li Zhou	Chinese Geology University (Wuhan), China

Sponsorship Group Co-chairs

Xin Wang	Communication University of China, China
Xiaojing Liang	NetEase Cloud Music, China
Yidan Zhu	Beijing Acoustics Academic Society, China

Publicity Group Co-chairs

Shufei Duan	Taiyuan University of Technology, China
Tianyi Zhang	Shanghai University Shanghai Film Academy, China

Equipment Debugging Technical Instruction Group Co-chairs

Xin Wang	Communication University of China, China
Tianyi Zhang	Shanghai University Shanghai Film Academy, China

Local Conference Affairs Group

Yanping Wang	North China University, China
Chao Lv	North China University, China
Weijia Liu	North China University, China
Juan Yang	North China University, China

Finance Section Co-chairs

Zuoliang Ning	Shanghai Computer Music Association, China
Zhongzhe Xiao	Suzhou University, China

Local Academic Exchange Section Co-chairs

Yanping Wang	North China University, China
Haifeng Li	Harbin Institute of Technology, China
Zuoliang Ning	Shanghai Computer Music Association, China

Sound and Music Apparatus Exhibition Committee Co-chairs

Zijin Li	China Conservatory of Music, China
Xiaomo Bai	Sichuan Conservatory, China
Wenlin Ban	China Conservatory of Music, China
Zuoliang Ning	Shanghai Computer Music Association, China
Mingchi Chan	Xinghai Conservatory of Music, China
Tianyi Zhang	Shanghai University Shanghai Film Academy, China
Rongfeng Li	Beijing University of Posts and Telecommunications, China
Chuyi Cui	Longy School of Music of Bard College, USA

Data Challenge Committee Co-chairs

Ru Zhang	Beijing University of Posts and Telecommunications, China
George Fazekas	Queen Mary University of London, UK
Zijin Li	China Conservatory, China
Yidan Zhu	Beijing Acoustics Academic Society, China
Wei Zhou	Beijing Zhongwen Law Firm, China
Shengchen Li	Beijing University of Posts and Telecommunications, China

The 4th Chinese Traditional Music Technology Session (CTMTS Co-chairs)

Rongfeng Li	Beijing University of Posts and Telecommunications, China
Xin Wang	Communication University of China, China
Jingyu Liu	Communication University of China, China

The 3rd General Audio-based Computer Audition Special Session Co-chairs

Ruohua Zhou	Beijing Construction University, China
Gang Tang	Beijing Chemical industry University, China
Maoshen Jia	Beijing University of Technology, China

The 1st Artificial Audition, Artistic Voice, Pathologic Voice Session Co-chairs

Lan Tian	Shan Dong University, China
Liyan Han	Central Conservatory of Music, China

Contents

Computational Musicology

Chorus Detection Using Music Structure Analysis

Zhengyu Cao[1], Yongwei Gao[1], and Wei Li[1,2](✉)

[1] School of Computer Science, Fudan University, Shanghai 201203, China
{zycao18,ywgao16,weili-fudan}@fudan.edu.cn
[2] Shanghai Key Laboratory of Intelligent Information Processing, Fudan University, Shanghai 201203, China

Abstract. This paper describes a novel chorus detection method based on extracting the functional structure of music from its self-similarity matrix. An existing similarity measure was enhanced firstly by using a key-shift invariant distance and by introducing a chroma-like pitch feature that exploits melody extraction results of the music. The repeated sections in the audio were extracted using a graph-based algorithm and clustering-merging method assuming transitivity of similarity then. Finally, a classifier to detect the chorus from the repeated sections was trained. The evaluation results show that our method is comparable with the state-of-the-art algorithms on both multiple and single chorus section detection tasks.

Keywords: Chorus detection · Music structure · Graph algorithm

1 Introduction

Music tend to be structured audio as described in [11], composed of repeating patterns/segments in hierarchies, from repeating phrases to sections. Among them, the longest repeating segments which correspond to sections or functional parts in the song are especially useful. For popular music, the basic song structure consists of an intro, verse, bridge, chorus and outro section. Chorus sections as the most representative parts of pop music are of special interest in many music-related applications, like auto music clipping, music thumbnailing, preview, retrieval and recommendation. For example, with the rapid growth of short-video services, the catchiest part of the music was preferred for making the videos. The service provider usually have large repositories of digital music clips which means clipping and choosing the chorus section manually is difficult, auto-clipping solves the problem.

2 Previous Work

To catch chorus sections, approaches based on music structure analysis were pervasively adopted, though there exist methods like [9] which directly estimate

X. Shao et al. (Eds.): CSMT 2020, LNEE 761, pp. 3–17, 2021.
https://doi.org/10.1007/978-981-16-1649-5_1

chorus sections from music audio. Self-similarity matrix (SSM) is the key component in many structure analysis algorithms [4,6,7,10,21,24].

One challenge faced in SSM based music structure analysis methods is how to extract meaningful segments from a raw SSM, which involves SSM enhancement and analysis. Various methods have been proposed to enhance the SSM, like matrix fusion technique from [2] used by [24], non-negative matrix factorisation (NMF) based methods [4,12] and methods [10] using augmentation of transposition and tempo invariance. As for extracting segments from the SSM, in [24], a spectral clustering method based on eigenvector decomposition of Laplacian matrix of the SSM was used to group the frames; in [4], a checkerboard kernel was applied to the SSM to generate a novelty curve, then peaks in the novelty curve were detected as segment boundaries; in [6] and [7], lines from the SSM diagonals were extracted first and merged using various hand-craft rules to form segments.

In [4], transitivity (which means if A and B are similar, and A and C are similar, then A and C should be similar) was enforced to the output music structure on the last step, the proposed method pushed it further: transitivity of similarity was considered at the first place, and this constraint was kept throughout the following steps. We proposed a novel graph-based algorithm to extract repeating segments from the SSM using a clustering-merging method. The clustering step can be seen as a repetition based method, comparing to the stripe detection approach used in previous repetition based methods like [7,14,18], the proposed method focus on detecting repeating patterns of smaller size but more repetitions, thus involve less effort integrating the repeating patterns and better reflects the repetition in music.

Melody extraction results were introduced to enhance the SSM in the proposed method, the reason is twofold. On the one hand, the feature-level similarity fusion in [24] draws good results in SSM enhancement while it supports arbitrarily many features as input so that new features could be added. On the other hand, the results of melody extraction algorithms has been greatly improved from salience-based approaches [22] to data-driven approaches like [1,3,13].

Heuristic methods were heavily used for chorus detection in previous works [6,7,17], focusing on distinguish chorus sections by repetition counts, durations and other features. Since the number of features could be large, and annotated datasets were available, supervised methods were preferred, in [25], a random forest classifier were used to detect chorus segments. We adopted the data-driven method, and combined melody extraction results into the features for chorus classifying.

Experiment on RWC Pop Database [8] shows our method is better comparing to the music thumbnailing algorithm in [9] on single section chorus detection, and has comparable performance with the best of 5 structure analysis algorithms mentioned in [19]. For reproducibility, the proposed algorithm and evaluation code is available on https://github.com/beantowel/chorus-from-music-structure.

3 Method

Figure 1 demonstrates the process of the proposed method. Firstly, acoustic features as pitch chroma, MFCC, chroma and tempogram were calculated from the input music recording. Then self-similarity matrices were generated on these features and fused into one. Low-level patterns were extracted by graph algorithms assuming transitivity of similarity and merged to form top-level structures. In the end, a classifier learns from the training data to detect chorus sections and makes predictions on structural information and melody features of the input sections.

Figure 2 shows the results from the pipeline. The fused SSM were plotted in the upper-left subfigure, the ground truth chorus sections and detected chorus sections were represented by the green stripes in the upper half and lower half of the box. The upper-right subfigure shows the ground truth structure annotations of the music, the green squares were verse sections and the blue squares were the chorus sections. The lower-right and lower-left subfigure shows the extracted low level and top-level structure of the music, different colors were used only to identify different repeating patterns.

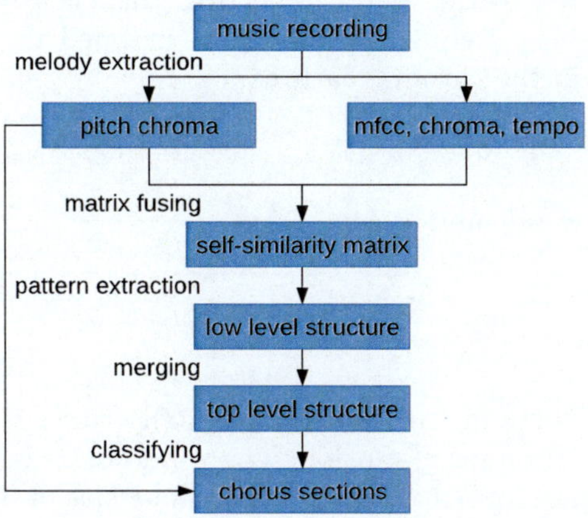

Fig. 1. Overview of the proposed method

3.1 Pitch Chroma Feature

The proposed method adopts the melody extraction result from [13] which is a melody line: a sequence of estimated fundamental frequency $\{f_0, f_1, f_2, ...\}$ corresponding to each timestamp $\{t_0, t_1, t_2, ...\}$. Though the algorithm has state-of-the-art results, wrong estimations in the output make the raw sequence not suitable for measuring similarity directly. Inspired by chroma feature, a pitch chroma feature vector is derived from the fundamental frequency sequence which is robust to the errors.

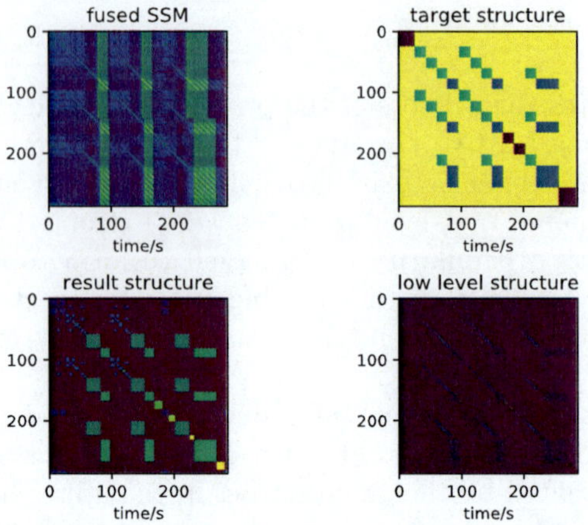

Fig. 2. Results for song 'Dream magic' from RWC Pop database

For a given window of frequencies $\{f_i, \ldots, f_j\}$ and a given number of pitch classes N_{class}, frequency values belonging to each pitch class were counted as pc_i, comprising a feature vector reflecting the occurrence of the pitches $\left[pc_0 \ldots pc_{N_{class}-1}\right]$. The frequency value f is mapped to its pitch class in a similar way to that in the chroma feature:

$$pitchClass(f) = N_{class} \log_2(f) \bmod N_{class} \tag{1}$$

the occurrences were counted as:

$$pc_k = \sum_{l=i}^{j} [pitchClass(f_l) = k] \tag{2}$$

In the proposed method, the number of pitch classes is set to $N_{class} = 24$ which gives the vector a finer resolution. The window size is $0.1 * 10$ s long, while the SSM used in the proposed method has a frame size of 0.23 s.

3.2 Key-Shift Invariant Distance

Modulation is the change of tonality, modulated sections are considered as the same pattern in the proposed structure analysis method. To deal with the key change in modulated sections, a key-shift invariant distance is used as the similarity measure for chroma and pitch chroma feature vectors.

For two feature vectors denoted as:

$$\mathbf{x} = \left[x_0 \ldots x_{n-1}\right], \mathbf{y} = \left[y_0 \ldots y_{n-1}\right] \tag{3}$$

n cosine similarity values is calculated by rolling the element in vector $\mathbf{y}$ by an offset of $i = 0, \ldots, n-1$ and evaluating the similarity between $\mathbf{x}$ and

$[y_{0+i} \cdots y_{n-1+i \bmod n}]$. The maximum similarity value, or the minimum distance among them is presented as the key-shift invariant distance of the two vectors. For example, the chroma feature has 12 pitch classes, then the distance is invariant to key changes in semitones.

3.3 Repeating Pattern Extraction

Low Level Pattern Extraction. Using the modified version of the algorithm from [24] with key-shift invariant distance, a self-similarity matrix is calculated by fusing SSMs of Mel-frequency cepstral coefficients (MFCC), chroma, pitch chroma and tempogram feature vectors. The fused SSM is 'cleaner' where stripe patterns corresponding to repeating segments were highlighted.

The fused SSM was binarized according to a threshold of $\exp(-5)$, values lower than the threshold were set to 0 while the rest were set to 1. The proposed method takes the binarized matrix as the adjacency matrix of the self-similarity graph (SSG) g_{ss} where vertices represents audio frames and edges represents the similarity relation between the frames.

Low-level patterns, or short repeating segments, were captured first. Similar frames forms a fully connected subgraph, or a clique in the SSG. By extracting cliques from the SSG, redundant or wrong edges representing a similarity relation were removed. However, cliques may overlap in the SSG, as there are noise/errors in the generated graph breaking the transitivity of similarity. To deal with the noise and find the repeating segments, the clique with maximum size was iteratively extracted from the graph as described in Algorithm 1, once a clique was extracted, the vertices in that clique were removed from the graph. This step requires to find all possible cliques in the SSG which is time-consuming when the graph is big, so literally only the first 10000 cliques found were used for selecting the largest clique, with less cliques to select, results will be slightly worse while decreasing processing time.

Figure 3 shows 2 example cases of extracting cliques from an undirected graph, cliques were enclosed in red-dotted circles, in the right subfigure, there are possible cliques that overlap with each other.

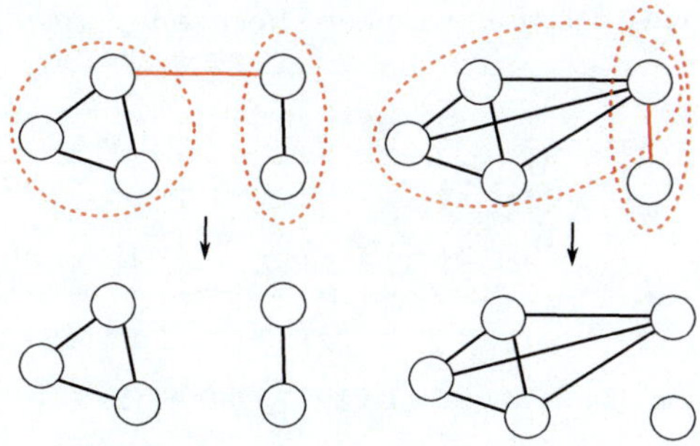

Fig. 3. Extracting cliques from undirected graph

Algorithm 1. Extract cliques from graph g_{ss}

Require: g_{ss} is undirected graph
Ensure: $C = \{c, \dots\}$ are non-overlap cliques in g_{ss}
  $C \leftarrow \{\}$
  while g_{ss} is not empty **do**
    $x \leftarrow findCliques(g_{ss})$
    $c \leftarrow maxSizeClique(x)$
    insert c into C
    remove c from g_{ss}
  end while
  return C

The extracted cliques were treated as low level patterns, each represents a group of repeating segments. For clarity, unique numbers can be assigned to the cliques, then the music structure will be represented by a sequence of label numbers for audio frames. For visualization, a labeled SSM can be constructed using the label sequence. The three representations, cliques, label sequence and labeled SSM, are equivalent data structures since they transform to each other freely. For example, cliques $C = \{(0,1),(2,3)\}$, label sequence $S = \{1,1,2,2\}$ and labeled SSM $M_{label} = \begin{bmatrix} 1\,1\,0\,0 \\ 1\,1\,0\,0 \\ 0\,0\,2\,2 \\ 0\,0\,2\,2 \end{bmatrix}$ all refer to the same structure.

Cliques Merging. To get functional level structure of music, the method merged the original extracted cliques to form larger repeating segments. Consider a music piece with structure ABA for example, the low level structure would look like $a_0a_0a_1a_2b_0b_1a_0a_0a_1a_2$ and the cliques extracted would be $\{(0,1,6,7),(2,8),(3,9),(4),(5)\}$. The target structure ABA, however, yields a target list of clique as $C = \{(0,1,2,3,6,7,8,9),(4,5)\}$. The principle is, if two cliques are sequentially adjacent, like a_0 and a_1 whose clique representation is $(0,1,6,7)$ and $(2,8)$, they can be merged into the same larger clique, as depicted in Fig. 4, where sequentially adjacent frames were connected by red lines. The transitivity of similarity relation were kept between merged repeating segments.

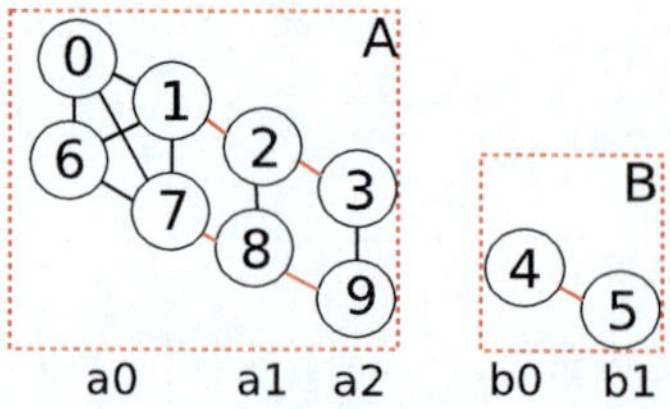

Fig. 4. Merging cliques into larger repeating segments

To decide whether two cliques were sequentially adjacent, we introduce the notion of 'ends' of the clique: it's composed of heads and tails (endpoints excluded) of the consecutive segments in the clique. For example, $(0, 6)$ and $(1+1, 7+1)$ are the heads and tails of clique a_0. Given two clique c_i, c_j where the minimum frame number satisfies $min(c_i) < min(c_j)$, tails of the former clique $tails(c_i)$ and heads of the latter clique $heads(c_j)$ were compared to tell if they met the adjacent condition.

Ideally, if the two cliques were adjacent, the heads and the tails should match as $tails(c_i) = heads(c_j)$, but to tolerate deviations the method uses a predicate which is a conjunction of:

- repeating counts restriction: difference between the numbers of consecutive segments in c_i and c_j is within D_{block}, which means for the lengths of the heads and tails, condition

$$|len(heads(c_i)) - len(tails(c_j))| < D_{block} \tag{4}$$

  is satisfied.
- distance restriction: distance between $tails(c_i), heads(c_j)$ is within D_{adj}, which means for most (except for at most D_{block} items) of the items $x \in tails(c_i)$ or $y \in heads(c_j)$, condition

$$\min_{\forall y \in heads(c_j)} |x - y| < D_{adj} \tag{5}$$

  or

$$\min_{\forall x \in tails(c_i)} |y - x| < D_{adj} \tag{6}$$

  is satisfied.

To merge original cliques into larger cliques. An adjacency matrix of the cliques M_{clique} whose items are $m_{i,j}$ is constructed by evaluating the predicate mentioned above for cliques c_i and c_j where $i < j$. According to the principle 'sequentially adjacent cliques be merged into the same larger clique', let g_{clique} be a graph with adjacency matrix M_{clique}, cliques in the same connected components were to be merged into one as described in Algorithm 2.

Smoothing and Adaptive Merging. To reduce noise/errors in the final output of structure analysis, a median filter with window size K_{window} is applied to the label sequence representation of the structure which smooth the output and keeps large repeating segments in one piece.

The parameter D_{block} is crucial for controlling the hierarchy of the output structure, with bigger D_{block} the criteria in Sect. 3.3 is more tolerant and results in larger repeating segments, with smaller D_{block}, the criteria is more strict and results in lower-level patterns. There is no optimal value of D_{block} for every song and its SSM, thus an adaptive merging method was adopted. From multiple merging outputs with $D_{block} \in \{0, \ldots, 2\}$ and $K_{window} \in \{23, 37, 47\}$, result whose count of cliques is greater than 3 and of minimum 'error' is selected.

Algorithm 2. Merge cliques

Require: $C = \{c, \dots\}$ are cliques
Ensure: $C_{merge} = \{c', \dots\}$ are largest possible merged cliques
  $M_{clique} = [m_{i,j}]$
  for $c_i, c_j \in C$ **do**
    $m_{i,j} \leftarrow isAdjacent(c_i, c_j)$
    $m_{j,i} \leftarrow m_{i,j}$
  end for
  $C_{merge} \leftarrow \{\}$
  for $x \in components(M_{clique})$ **do**
    $c' \leftarrow ()$
    for $c \in x$ **do**
      add c into c'
    end for
    insert c' into C_{merge}
  end for
  return C_{merge}

The error of clique merging process is modeled by comparing the labeled SSMs of the original clique and that of the merged clique. Empirically, a good merging result is close to the original cliques, thus we use an error function composed of two terms: false negative rate and false positive rate. Given the original cliques C and merged cliques C_{merge}, their labeled self-similarity matrix representations $m_c = M_{label}(C)$ and $m_{c'} = M_{label}(C_{merge})$ were compared. The error function is:

$$Error(C, C_{merge}) = \alpha E_{Neg} + max(\beta E_{Pos} - 0.1, 0) \tag{7}$$

where $E_{Neg} = sum(m_{c'} = 0 \wedge m_c \neq 0)$ is the number of false negatives and $E_{Pos} = sum(m_{c'} \neq 0 \wedge m_c = 0)$ is the number of false positives, coefficients $\alpha = \frac{1}{sum(m_c \neq 0)}$ and $\beta = \frac{1}{sum(m_c \neq 0)}$ were used to normalize the importance of the terms as both type of errors are considered equally important. Good merged cliques should cover the original cliques, thus the false positives were inevitable and always greater than 0, so the minus-then-max function clips the false positive rate lower then 0.1.

3.4 Chorus Detection

Based on the music structure analysis results, the chorus detection task is just of choosing the right repeating segments as the chorus. The proposed method uses a random forest classifier to learn which cliques are the chorus sections. The chorus sections tend to have different acoustic features to other sections and specific positions in the song, thus acoustic and structural features of the cliques $C = \{c, \dots\}$ were provided to train the classifier, the features used were listed below:

- *clique duration*: duration occupied by the clique normalized by duration of the song.
- *voicing rate*: the ratio of the number of voicing frames to that of all frames in the clique by counting non-zero frequencies in the melody line within the clique.
- *melody median*, *minimum* and *maximum*: median, minimum and maximum value of the frequencies in the melody line within the clique.
- *clique head* and *last clique head*: the smallest and the biggest frame number in the heads of the clique $heads(c)$.
- *segments count*: the number of consecutive segments in the clique by measuring the size of $heads(c)$.

Apart from the structural features like *clique head*, we added more features to expose the positional/structural information of the cliques. Based on the 8 features mentioned above, in 3 ways additional features were generated:

- *ranking*: features of the cliques in a music recording were ranked by sorting their values, the ranking numbers were used as additional features.
- *normalizing*: features were normalized by the maximum value from the cliques in a music recording to generate additional features.
- *stacking*: cliques were sorted by occurrence (their minimum frame number), then features of a clique's predecessor and successor were copied and added as additional features.

In the training phase, the proposed structure analysis algorithm was first applied to the music recordings in the training set to get repeating segments, then each clique was compared with the ground truth annotation to decide whether to label it as a chorus section or not. The comparison is done by measuring the overlap ratio between the clique and the ground truth chorus sections, given the length of the clique l_{clique}, the length of the chorus section and of overlap part $l_{chorus}, l_{overlap}$, two metrics can be calculated as:

- precision: $p = \frac{l_{overlap}}{l_{clique}}$
- recall: $r = \frac{l_{overlap}}{l_{chorus}}$

Cliques with precision $p > 50\%$ and recall $r > 10\%$ were labeled as chorus sections, the others were labeled as non-chorus sections. For each clique, an 8 dimensional feature vector and an $8 * 4 = 32$ dimensional additional feature vector were calculated, the data were used to train a random forest classifier with 1000 decision trees.

In the prediction phase, features of the cliques extracted by the structure analysis algorithm were fed to the classifier, the output were directly used as the result of chorus detection.

4 Evaluation

4.1 Database and Metrics

The RWC Pop Database [8] used for evaluation contains 100 popular songs, each with one functional structure annotation file. To train the classifier for chorus detection, the dataset was randomly split into a training set and a validation set which constitutes of 70 songs and 30 songs respectively. The evaluation do not distinguish between different chorus sections like 'chorus A' and 'chorus B' which were used in the annotation. Chorus detection performance was measured by overlap-based metrics like in Sect. 3.4 and [7]:

- precision: $P = \frac{\text{total length of correctly detected chorus sections}}{\text{total length of detected chorus sections}}$
- recall: $R = \frac{\text{total length of correctly detected chorus sections}}{\text{total length of correct chorus sections}}$
- f-measure: $F = \frac{2RP}{R+P}$

For comparison with algorithm from [9] which outputs single chorus section, a modified version of the above metrics was used. Only nearest ground truth chorus section was considered when measuring the output chorus section. If there were multiple output chorus section, the length of distinct nearest correct chorus sections were summed up as $L_{nearest\ chorus}$. Given the total length of correctly detected nearest chorus sections $L_{nearest\ overlap}$, the modified metrics were denoted as:

- precision-single: $P_{single} = \frac{L_{nearest\ overlap}}{\text{total length of detected chorus sections}}$
- recall-single: $R_{single} = \frac{L_{nearest\ overlap}}{L_{nearest\ chorus}}$
- f-measure-single: $F_{single} = \frac{2R_{single}P_{single}}{R_{single}+P_{single}}$

The modified metrics are suitable for algorithms detecting a single chorus section and are compatible with algorithms detecting multiple chorus sections, for the latter case, the precision and recall can be viewed as an averaged score for the multiple detected chorus sections.

4.2 Reference Methods

The proposed method was denoted as 'seqRecur', for better comparison with [9], a modified version of the method denoted as 'seqRecurS' which has the same single section output format were also evaluated. The fixed-length single chorus section which covers most of the chorus sections predicted by the proposed method was selected as the output of 'seqRecurS'.

Apart from the proposed method, we evaluated 6 reproducible algorithms, denoted as 'highlighter', 'scluster', 'sf', 'olda', 'cnmf' and 'foote' [5,9,15,16,20,23]. MSAF [19] implementation of the latter 5 structure analysis algorithms from https://github.com/urinieto/msaf were used. To evaluate the performance of structure analysis algorithms on chorus detection task, extra steps were taken.

For label algorithms from MSAF (scluster, cnmf) which outputs the music structure via labeled sections, the chorus detection method as in Sect. 3.4 can be applied. For boundary algorithms from MSAF (sf, olda, foote) which split a music recording into sections, since the output has no recurrent music structure but only boundaries, the output sections was labeled by maximizing similarity on the SSM used in Sect. 3.3, then the same chorus section classifier can be applied on these sections.

To assess the effect of the chorus detection classifier independently, ground truth music structure were provided in the training and prediction phase of the chorus classifier, the result was denoted as 'gt'. To assess the effect of the structure analysis algorithm independently, the output sections can be assigned labels to achieve the highest possible precision, i.e. the section was labeled as chorus if more than 50% of its length overlap with ground truth chorus sections, its results were denoted by plus sign suffix like 'scluster+' as it represents the upper bound chorus detection precision of a structure analysis method.

4.3 Results

The average precision, recall and f-measure for songs in the validation set were listed in Table 2. The violin plot which shows the minimum, maximum and average value of F, F_{single} for songs in the validation set were shown in Fig. 5. The proposed method 'seqRecur' was the best on R, F among other structure analysis algorithms, though its upper bound performance 'seqRecur+' was worse than that of 'olda+'. The modified proposed method 'seqRecurS' were comparable on F_{single} than 'highlighter', which was designed for detecting a single chorus section.

The high scores of 'gt' shows that the chorus detection classifier was capable of learning from the human-labeled ground truth structure annotations. The performance decrease from results of highest possible precision 'X+' to its correspondence 'X' using the classifier to detect chorus sections shows that the output of existing structure analysis algorithms didn't fit chorus detection task well, one possible reason is that the output structure lacks consistency, making it difficult to learn to distinguish the chorus from other functional sections.

Table 1. Reference method categories

	Ground truth structure	Calculated structure
Ground truth chorus	–	X+
Calculated chorus	gt	X

4.4 Ablation Study

To verify the effect of enhancing SSM by introducing the pitch chroma feature, we conduct an ablation study by removing the pitch chroma feature used in the SSM fusion step. With the same parameter settings, the evaluation results were listed in Table 3. By utilizing the melody extraction results with the pitch chroma feature, the performance of the chorus detection system increased by 2% in f-measure for method 'seqRecur'.

Table 2. Average results on validation set. Depending on whether the two stages of the algorithms: structure analysis and chorus detection have used ground truth results, the reference methods can be divided into 3 categories as described in Table 1.

algo	P	R	F	P_{single}	R_{single}	F_{single}
seqRecur	0.8050	**0.7688**	**0.7726**	0.7957	0.7771	0.7701
seqRecurS	0.8533	0.3317	0.4680	0.8198	0.7212	0.7508
highlighter	**0.8820**	0.3354	0.4762	**0.8571**	0.7784	0.7910
scluster	0.6772	0.6528	0.6038	0.6436	0.6912	0.6324
sf	0.7889	0.7068	0.6906	0.7446	0.8303	0.7423
olda	0.7793	0.7615	0.7313	0.7304	0.8675	0.7571
foote	0.8368	0.6812	0.7030	0.7886	**0.8738**	**0.8068**
seqRecur+	0.8573	0.8031	0.8169	**0.8436**	0.8155	0.8168
scluster+	0.8545	0.8972	0.8672	0.8107	0.9227	0.8470
sf+	0.8108	0.8841	0.8323	0.7624	0.9221	0.8141
olda+	0.8683	**0.9344**	**0.8940**	0.8221	**0.9548**	**0.8762**
cnmf	0.5510	0.6482	0.5714	0.5262	0.6966	0.5805
cnmf+	0.7929	0.8594	0.8078	0.7457	0.9063	0.8065
foote+	**0.8717**	0.8684	0.8621	0.8270	0.8898	0.8452
gt	0.9395	0.9460	0.9423	0.9253	0.9463	0.9328

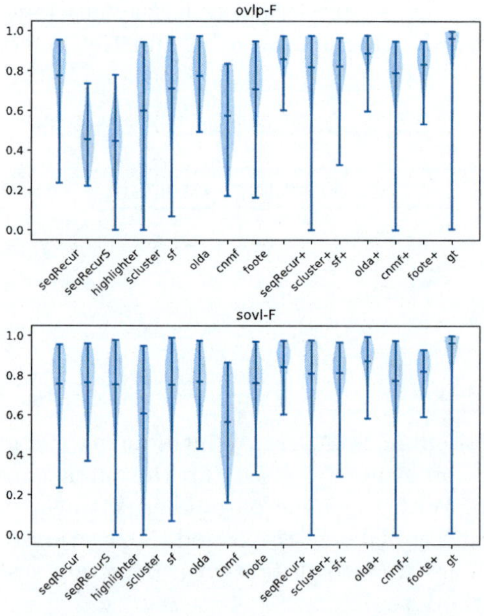

Fig. 5. Distribution of F (ovlp-F) and F_{single} (sovl-F) on validation set

Table 3. Performance increase when pitch chroma feature was used to enhance SSM

algo	ovlp-P	ovlp-R	ovlp-F	sovl-P	sovl-R	sovl-F
seqRecur	−0.98%	5.23%	2.31%	−1.48%	3.85%	1.02%
seqRecurS	−6.52%	−1.69%	−2.76%	−6.82%	−3.58%	−5.26%
seqRecur+	2.21%	−5.31%	−2.00%	1.74%	−4.13%	−1.44%

5 Conclusion

This paper proposed a chorus detection method based on music structure analysis results. To better compute the music similarity, we enhanced an existing similarity fusing method by introducing a new feature which exploits melody extraction algorithms and a key-shift invariant distance to deal with the key changes. A novel structure analysis method using graph algorithms and a chorus detection method using supervised learning was proposed.

The chorus detection method were applied to the output of the proposed structure analysis algorithm and the other 5 state-of-the-art algorithms. Evaluation results shows the method was comparable with the state-of-the-arts algorithms on both multiple and single chorus section detection tasks. The adapted structure analysis methods using part of the proposed method to detect chorus sections also reach high performance.

Results shows utilizing music structure analysis and melody extraction algorithms for chorus detection was viable and competitive. However, the performance was still not satisfactory comparing to the upper bound. Two reasons lie behind this: the structure analysis algorithms were not good enough, and the ambiguity of what ‘chorus’ means since the arrangement of songs varies and the functional sections were annotated by humans.

Ackowlegement. This work was supported in part by National Key R&D Program of China (2019YFC1711800), NSFC (61671156).

References

1. Bittner, R.M., McFee, B., Salamon, J., Li, P., Bello, J.P.: Deep salience representations for F0 estimation in polyphonic music. In: The 18th International Society for Music Information Retrieval Conference, Suzhou, China, pp. 63–70 (2017)
2. Wang, B., Jiang, J., Wang, W., Zhou, Z.-H., Tu, Z.: Unsupervised metric fusion by cross diffusion. In: 2012 IEEE Conference on Computer Vision and Pattern Recognition, IEEE, Providence, RI, pp. 2997–3004 (2012). https://doi.org/10.1109/CVPR.2012.6248029, http://ieeexplore.ieee.org/document/6248029/
3. Chen, M.T., Li, B.J., Chi, T.S.: CNN based two-stage multi-resolution end-to-end model for singing melody extraction. In: ICASSP 2019 - 2019 IEEE International Conference on Acoustics, Speech and Signal Processing (ICASSP), Brighton, United Kingdom, pp. 1005–1009. IEEE (2019). https://doi.org/10.1109/ICASSP.2019.8683630, https://ieeexplore.ieee.org/document/8683630/

4. Cheng, T., Smith, J.B.L., Goto, M.: Music structure boundary detection and labelling by a deconvolution of path-enhanced self-similarity matrix. In: 2018 IEEE International Conference on Acoustics, Speech and Signal Processing (ICASSP), Calgary, AB, pp. 106–110. IEEE (2018). https://doi.org/10.1109/ICASSP.2018.8461319, https://ieeexplore.ieee.org/document/8461319/
5. Foote, J.: Automatic audio segmentation using a measure of audio novelty. In: 2000 IEEE International Conference on Multimedia and Expo. ICME2000. Proceedings. Latest Advances in the Fast Changing World of Multimedia (Cat. No.00TH8532), New York, NY, USA, vol. 1, pp. 452–455. IEEE (2000). https://doi.org/10.1109/ICME.2000.869637, http://ieeexplore.ieee.org/document/869637/
6. Gao, S., Li, H.: Popular song summarization using chorus section detection from audio signal. In: IEEE 17th International Workshop on Multimedia Signal Processing (MMSP), Xiamen, China, pp. 1–6. IEEE (2015). https://doi.org/10.1109/MMSP.2015.7340798, http://ieeexplore.ieee.org/document/7340798/
7. Goto, M.: A chorus section detection method for musical audio signals and its application to a music listening station. IEEE Trans. Audio Speech Lang. Process. **14**(5), 1783–1794 (2006). https://doi.org/10.1109/TSA.2005.863204, http://ieeexplore.ieee.org/document/1677997/
8. Goto, M., Hashiguchi, H., Nishimura, T., Oka, R.: RWC music database: popular, classical and jazz music databases. In: ISMIR, Paris, France, vol. 2, pp. 287–288 (2002)
9. Huang, Y.S., Chou, S.Y., Yang, Y.H.: Pop music highlighter: marking the emotion keypoints. Trans. Int. Soc. Music Inf. Retrieval **1**(1), 68–78 (2018). https://doi.org/10.5334/tismir.14, http://transactions.ismir.net/articles/10.5334/tismir.14/
10. Jiang, N., Muller, M.: Estimating double thumbnails for music recordings. In: 2015 IEEE International Conference on Acoustics, Speech and Signal Processing (ICASSP), South Brisbane, Queensland, Australia, pp. 146–150. IEEE (2015). https://doi.org/10.1109/ICASSP.2015.7177949, http://ieeexplore.ieee.org/document/7177949/
11. Jun, S., Rho, S., Hwang, E.: Music structure analysis using self-similarity matrix and two-stage categorization. Multimedia Tools Appl. **74**(1), 287–302 (2015). https://doi.org/10.1007/s11042-013-1761-9
12. Kaiser, F., Sikora, T.: Music structure discovery in popular music using non-negative matrix factorization. In: The 11th International Society for Music Information Retrieval Conference, Utrecht, Netherlands, p. 6 (2010)
13. Kum, S., Nam, J.: Joint detection and classification of singing voice melody using convolutional recurrent neural networks. Appl. Sci. **9**(7), 1324 (2019). https://doi.org/10.3390/app9071324, https://www.mdpi.com/2076-3417/9/7/1324
14. Lu, L., Wang, M., Zhang, H.J.: Repeating pattern discovery and structure analysis from acoustic music data. In: Proceedings of the 6th ACM SIGMM international workshop on Multimedia information retrieval - MIR 2004, New York, NY, USA, p. 275. ACM Press (2004). https://doi.org/10.1145/1026711.1026756, http://portal.acm.org/citation.cfm?doid=1026711.1026756
15. McFee, B., Ellis, D.P.: Learning to segment songs with ordinal linear discriminant analysis. In: 2014 IEEE International Conference on Acoustics, Speech and Signal Processing (ICASSP), Florence, Italy, pp. 5197–5201. IEEE (2014). https://doi.org/10.1109/ICASSP.2014.6854594, http://ieeexplore.ieee.org/document/6854594/
16. McFee, B., Ellis, D.P.W.: Analyzing song structure with spectral clustering. In: The 15th International Society for Music Information Retrieval Conference, Taipei, Taiwan, p. 6 (2014)

17. Mildner, V., Klenner, P., Kammeyer, K.D.: Chorus detection in songs of pop music. In: Proceedings of ESSV, Universität Karlsruhe, Karlsruhe, p. 8 (2003)
18. Müller, M., Kurth, F.: Towards structural analysis of audio recordings in the presence of musical variations. EURASIP J. Adv. Sig. Process. **2007**(1) (2006). https://doi.org/10.1155/2007/89686, https://asp-eurasipjournals.springeropen.com/articles/10.1155/2007/89686
19. Nieto, O., Bello, J.P.: Systematic exploration of computational music structure research. In: The 17th International Society for Music Information Retrieval Conference, New York City, USA, p. 7 (2016)
20. Nieto, O., Jehan, T.: Convex non-negative matrix factorization for automatic music structure identification. In: 2013 IEEE International Conference on Acoustics, Speech and Signal Processing, Vancouver, BC, Canada, pp. 236–240. IEEE (2013). https://doi.org/10.1109/ICASSP.2013.6637644, http://ieeexplore.ieee.org/document/6637644/
21. Paulus, J., Klapuri, A.: Audio-based music structure analysis. In: The 11th International Society for Music Information Retrieval Conference, Utrecht, Netherlands, p. 12 (2010). citation Key Alias: paulusAUDIOBASEDMUSICSTRUCTURE2010a
22. Salamon, J., Gómez, E.: Melody extraction from polyphonic music signals using pitch contour characteristics. IEEE Trans. Audio Speech Lang. Process. **20**(6), 1759–1770 (2012)
23. Serra, J., Muller, M., Grosche, P., Arcos, J.L.: Unsupervised detection of music boundaries by time series structure features. In: Twenty-Sixth Conference on Artificial Intelligence, Toronto, Ontario, Canada, p. 7 (2012)
24. Tralie, C.J., McFee, B.: Enhanced hierarchical music structure annotations via feature level similarity fusion. In: ICASSP 2019 - 2019 IEEE International Conference on Acoustics, Speech and Signal Processing (ICASSP), Brighton, United Kingdom, pp. 201–205. IEEE (2019). https://doi.org/10.1109/ICASSP.2019.8683492, https://ieeexplore.ieee.org/document/8683492/
25. Wu, F., Sun, S., Xue, W.: Automatic extraction of popular music ringtones based on music structure analysis. In: 2016 IEEE/ACIS 15th International Conference on Computer and Information Science (ICIS), Okayama, Japan, pp. 1–5. IEEE (2016). https://doi.org/10.1109/ICIS.2016.7550919, http://ieeexplore.ieee.org/document/7550919/

Deconstruct and Reconstruct Dizi Music of the Northern School and the Southern School

Yifan Xie and Rongfeng Li(✉)

Beijing Key Laboratory of Network System and Network Culture,
Beijing University of Posts and Telecommunications, Beijing 100876, China
{yifan.xie,lirongfeng}@bupt.edu.cn

Abstract. Today's research on Chinese music technology is mainly focused on three aspects: data collection, music deconstruction, and music reconstruction. In this paper, a general method is proposed to collect Chinese music in the form of numbered musical notation, and a Dizi dataset is collected using this method. Based on the collected Dizi dataset, we conduct research on the Dizi music styles of the Northern school and the Southern School. Characteristics include melody and playing techniques of the two different music styles are deconstructed. A reconstruction example, music style transfer which includes melody transfer and playing techniques transfer is given and audience evaluation is done to evaluate the reconstruction results.

Keywords: Dizi music · Deconstruction · Reconstruction

1 Introduction

With the continuous development of music technology, more and more researchers are devoted to the exploration of Chinese music technology. These studies mainly focus on three aspects: the collection of Chinese music datasets, the deconstruction of Chinese music, and the reconstruction of Chinese music. Collected datasets include two forms: audio and symbolic scores. Deconstruction refers to data mining from the collected datasets, to find some information about features from music. Reconstruction refers to the creation of new music, new musical forms, and so on.

Of course, there are still many areas worthy of improvement. The current research on Chinese music mainly has the following problems:

- In terms of data collection, there has been a relatively standardized and systematic collection method for audio, and there is also a certain scale of Chinese musical instrument database [12]. But in terms of the establishment of a symbolic score database, although some methods have been proposed before, there is still no standard, accurate and fast collection method for the numbered musical scores. For example, Optical Character Recognition

X. Shao et al. (Eds.): CSMT 2020, LNEE 761, pp. 18–28, 2021.
https://doi.org/10.1007/978-981-16-1649-5_2

technology is a fast method but could lead to many errors due to immaturity. The collection method proposed in [10] for Guqin is fast and accurate to some extent, but could not intuitively be represented in the form of numbered musical notation.
- In the reconstruction and deconstruction of music, the gap between music and computer science has caused two hands of problems. On the one hand, for researchers in computer science, much research stays at the stage of pure data analysis without an in-depth discussion of the meaning of the music. Simply migrating methods from other fields to Chinese music technology would not bring substantial progress to the research of Chinese music technology. On the other hand, for researchers in music, much research fails to make good use of the powerful tool of computer science.

In response to the above problems. Taking Dizi music as an example, we do the following work in this paper:

- we propose a general collection method to collect numbered musical scores by typing them using a self-made font. In this way, collected scores can be represented intuitively in the form of numbered musical notation. Then, these scores can be converted into staff easily using a written program. Using this method, we collect the first symbolic dataset of Dizi music. We also make public the dataset[1].
- Based on the Dizi dataset, we do data mining (deconstruction) include melody deconstruction and playing techniques deconstruction on the Dizi music styles. Playing techniques in Chinese music is much more important than in Western music. Through deconstruction, we not only lay the foundation for the later music reconstruction but also find some interesting phenomena.
- Based on the deconstruction results, we give an interesting reconstruction example, music style transfer, which includes both melody transfer and playing techniques transfer. Some audience tests are done to evaluate the transfer results.

The code used in this paper can be found online[2]. Besides, here is a brief introduction to styles of Dizi music, especially for the styles of the Northern school and the Southern school, which are seen as the research objects in this paper. In the 1950s, Dizi appeared on the historical stage of solo performance, and its performance styles consisted of the Southern school and the Northern school. Today, the Northern school and the Southern school are two main styles of Dizi music. The Northern school is characterized as more lively, and ornamentally technical with extensive use of different types of tricky fingering techniques and tonguing. The Northern school is mainly played by Bangdi (Check out Hong Kong Chinese Orchestra's introductory video to the Bangdi in https://www.youtube.com/watch?v=zJjfFqat_oA). By contrast, the Southern school is more melodic. It can display the soft features of the Jiangnan

[1] https://github.com/hrsoup/Dizi_Dataset.
[2] https://github.com/hrsoup/CSMT2020_Code.

region. The representative playing techniques of the Souther school include trills, upper acciaccatura, and so on. The Southern school is mainly played by Qudi (Check out Hong Kong Chinese Orchestra's introductory video to the Qudi in https://www.youtube.com/watch?v=HjU5ssXvYQA).

The rest of this paper is organized as follows. Related work is first shown in Sect. 2. Data preparation is described in Sect. 3. Section 4 and Sect. 5 introduce deconstruction and reconstruction, respectively. Conclusions and some future work are given in Sect. 6.

2 Related Work

The first part of related work is data collection. The collected data of Chinese music consists of two forms: audio and symbolic scores. In terms of audio database establishment, Liang et al. [12] built a database that includes Chinese musical instrument audio. Wang et al. [16] collected a Dizi audio music dataset. In terms of symbolic music scores collection, Li et al. [9] collected a Gongchepu (which is the Chinese traditional musical notation) dataset. Li et al. [10] collected a Guqin dataset.

The second part of related work is music deconstruction. Music deconstruction includes research about melody, music spectral characteristics, the correlation of different types of music genres, and so on. Wang et al. [17] did research about playing techniques recognization from the Dizi music spectrum. Yang et al. [21] did a quantitative study of vibrato to compare erhu music and violin music.

The third part of related work is music reconstruction. Music reconstruction includes music generation, music synthesis, and so on. Luo et al. [13] used methods of deep learning to generate Chinese folk songs with specific styles. Dai et al. [2] did music synthesis based on modeling of pipa playing techniques.

3 Data Preparation

3.1 Data Collection

Chinese musical instruments are usually recorded in the form of numbered musical notation. Therefore, we propose a general method to collect numbered musical scores. Although this method is applied in Dizi in this paper, it can be extended easily to other Chinese musical instruments which are also recorded in numbered musical notation. This method consists of three steps: making a new font, typing, and transformation.

First, we used the open-source software FontForge to make a new font. We call the new font DiziFont.ttf. The method of making DiziFont.ttf can be seen in Fig. 1. The left subfigure shows the overview of the font design. Some keys in the keyboard correspond to some symbols in the numbered musical notation. For example, the method of making lower acciaccatura is shown in the right

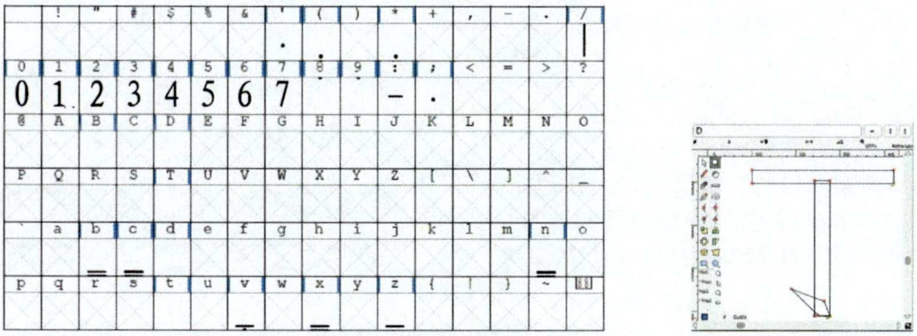

(a) The overview of the font design

(b) Making lower acciaccatura

Fig. 1. The method of making DiziFont.ttf

subfigure. The symbol of lower acciaccatura consists of three polygons of different shapes and sizes, and it corresponds to the capital letter D on the keyboard.

Second, we applied the font of DiziFont.ttf into Microsoft word to type in numbered musical notation. A piece of example can be seen in Fig. 2. From this figure, we can see that the digitized numbered musical notation looks the same as on paper, which is very intuitive. The paper sheet music we used comes from [11] and [20]. The original typing files are stored in Docx files.

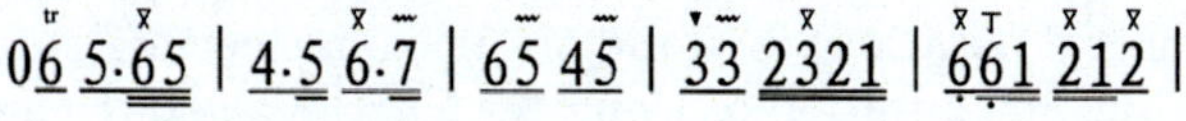

Fig. 2. A typing example

Third, we wrote a program using the music21 toolkit [1] to transform the Docx file into the MusicXML file. Although it is intuitive to record the numbered musical notation in the Docx file using our self-made font, it is not standardized. MusicXML file is not only a more standardized store form but also can be displayed in the form of staff using some software such as MuseScore. A transform example from the Docx file to the MusicXML file can be seen in Fig. 3. In practice, the playing technique symbols are too complicated which brings us great difficulties to process them. Therefore, we used an unintuitive but very simple way to record and process playing techniques. In this way, playing techniques are represented as lyrics to add to staff scores.

Using this above method, we collect a Dizi dataset both in Docx files and MusicXML files (which is also to say, numbered musical scores and staff scores).

3.2 Data Statistics

Up to now, we have recorded 28 Dizi songs, which include 19413 notes in total. The dataset is still being continuously expanded. In terms of style, these songs

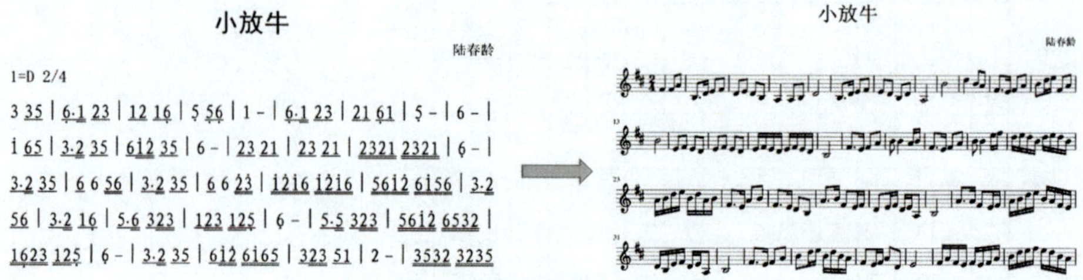

Fig. 3. A transform example

cover the Southern school, the Northern school, and so on. In this paper, we use the Northern school data and the Southern school data to analyze. The total number of notes used in this paper is 12925, where 7320 notes for the Northern school and 5605 notes for the Southern school.

3.3 Data Representation

In this paper, we focus on Dizi music of symbolic representation, so each note can be seen as a word just like in natural language processing. Each note consists of two features: the pitch and the duration. We use the chromatic scale to measure the pitch and quarter length to measure the duration. An example is shown in Fig. 4. It can be seen how the processing of symbolic music is related to natural language processing. It shows a note sequence (represented in the form of numbered musical notation) in C major. The quarter note Do in C major has the pitch of C4 and the duration of 1 quarter length. The pitch and the duration can be spliced together and expressed as C41, and the other notes are the same.

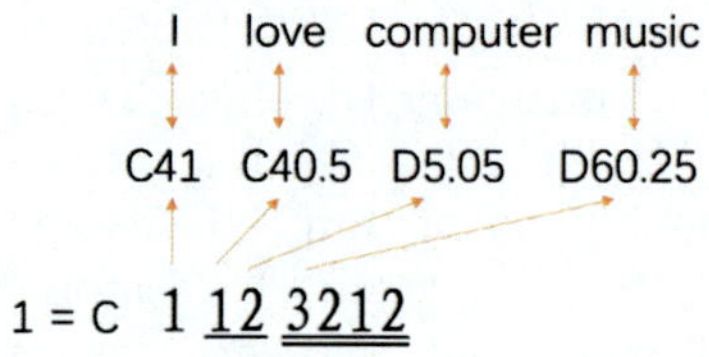

Fig. 4. A data representation example

4 Deconstruction

Deconstruction is used to find some information about features from music, which is done through data mining. For most kinds of music, the melody is quite an important feature, so we first did the melody deconstruction. For Dizi music, playing techniques are also important. Different styles have different representative playing techniques, so we then did the playing techniques deconstruction. In this paper, deconstruction is seen as a typical classification task.

4.1 Melody Deconstruction

Melody deconstruction is used to classify melodies of the Northern school and the Southern school, which can also be seen as a general text classification task. We first cut music into many pieces, each of whose length is 4 measures, then used data preprocessing techniques include Bag-of-Words, Term Frequency Inverse Document Frequency (TF-IDF) [15], Continuous Bag-of-Words (CBOW) [14] and Skip-Gram [14] to process these data, finally sent these preprocessed-data to some usual machine learning models include Support Vector Machines (SVM) [5], long-short-term memory (LSTM) [4] and Text Convolutional Neural Network (TextCNN) [7]. Experiments were done under 10-fold cross validation. Recall and F1-socre were used to evaluate experiment results.

The results are shown in Table 1. From this table, we can see that LSTM+TF-IDF and LSTM+Bag-of-Words get the relative best results. Besides, we find that the results of using LSTM are better than using TextCNN in total, which is an interesting result. In the short-text classification task, TextCNN has been proved to perform better than LSTM in many tasks, but the Dizi melody classification is not so. We think it is because that TextCNN can only do convolution operation during a small range, but LSTM can memory longer data. Compared with text data, music more depends on long memorized data. Grasping partial features not global features are difficult to recognize melody style.

Table 1. Melody deconstruction results

Model	Bag-of-Words		TF-IDF		CBOW		Skip-Gram	
	Recall	F1-score	Recall	F1-score	Recall	F1-score	Recall	F1-score
SVM	97.89	89.96	98.77	95.82	97.96	92.21	97.81	92.23
LSTM	98.41	97.45	98.10	97.74	95.35	94.16	93.87	94.72
CNN	89.66	89.28	96.11	95.40	93.76	88.71	87.83	82.39

4.2 Playing Techniques Deconstrction

Playing techniques deconstruction is seen as a special classification task, tagging task. The tagging task is discussed between the observation sequence and the state sequence. In this playing technique deconstruction, each kind of playing technique is seen as a state and each note is seen as an observation. We first cut music into pieces which one of whose length is 4 measures, then used random word embedding to preprocess data, finally sent these preprocessed-data to some tagging models. Experiments were done under 10-fold cross-validation in the dataset of the Northern school and the dataset of the Southern school, respectively. Accuracy and oov (out-of-vocabulary) accuracy were used to evaluate experimental results.

Besides, Special instructions are needed regarding the tagging model used. Besides these usual tagging models include Conditional Random Fields (CRF) [8], bidirectional LSTM (BILSTM) [3] and BILSTM with a CRF layer (BILSTM-CRF) [6], we also used the model proposed in [19], which combines a general tagging model and logic rules. As the general tagging model we used is BILSTM, we call this model BILSTM-RULES.

The total experiment results are shown in Table 2. We can see that BILSTM-CRF achieves the highest accuracy among two datasets of different styles, while BILSTM-RULES achieves the highest oov accuracy.

Table 2. Playing techniques deconstruction results. In this table, N represents the Northern school and S represents the Southern school.

Model	Accuracy		Oov accuracy	
	N	S	N	S
CRF	68.98	84.42	39.44	63.03
BILSTM	69.76	84.03	61.29	88.26
BILSTM-CRF	74.71	87.59	43.54	85.53
BISLTM-RULES	69.23	84.12	61.95	88.79

Besides, we find an interesting phenome, that is, not the same as the original data label does not mean it does not meet a certain style. Although composers from the same school can have different playing techniques tagging ways for the same music. For example, the first two subfigures in Fig. 5 show an excerpt of *Song of Soochow*. Xunfa Yu and Xianwei Jiang are both from the Southern school, but they tagging the music differently. In our playing techniques deconstruction, there are also some similar examples of happening. An example is shown in the last two subfigures in Fig. 5, it shows an excerpt of *Busy Delivering Harvest*. We can see that playing techniques are different between the original and the generated. But from the human perspective, the playing techniques generated by BILSTM-CRF still meet the original style, the Northern school. As tonguing (which is represented as a triangle in numbered musical notation) is a typical playing technique from the Northern school.

5 Reconstruction

After deconstruction, we can reconstruct new music using deconstruction results. In this paper, we use music style transfer as a reconstruction example, to show how to get new music from deconstruction results. There are two steps of reconstruction (music style transfer) in this paper: melody reconstruction (melody transfer) and playing techniques reconstruction (playing techniques transfer).

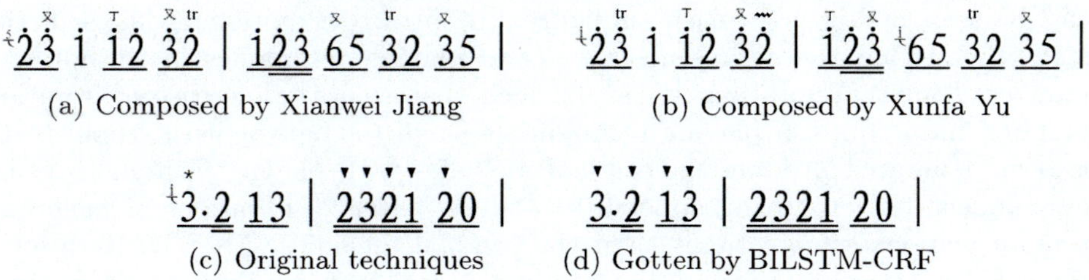

(a) Composed by Xianwei Jiang (b) Composed by Xunfa Yu

(c) Original techniques (d) Gotten by BILSTM-CRF

Fig. 5. Different techniques tags for the same notes

5.1 Melody Reconstruction

The first step is melody transfer. Given a piece of music with a specific style, then add some changes to this piece of music. The changes include the following four kinds:

- Changing a note one or two chromatic semitones higher than the old one.
- Changing a note one or two chromatic semitones lower than the old one.
- Splitting one note into several notes, these notes' duration include $(\frac{1}{2}, \frac{1}{2})$, $(\frac{1}{3}, \frac{2}{3})$, $(\frac{2}{3}, \frac{1}{3})$ and $(\frac{1}{3}, \frac{1}{3}, \frac{1}{3})$
- Joining two successive notes into one note whose duration is the sum of these two notes' duration.

Then, this piece of music with changes was sent to LSTM with the pre-processing technique of TF-IDF (which has been proved to perform well in Sect. 4.1). If the prediction label of classification is the same as its initial label (keeping the content of original music), and the probability of prediction is smaller than the old (It is closer to the style of target music), then these changes could be retained, else be dropped. The above process was repeated until the specified number of iterations is reached.

5.2 Playing Techniques Reconstruction

The second step is playing techniques transfer. We used models in Sect. 4.2 Which achieve the highest accuracy (BILSTM-CRF) and oov accuracy (BILSTM-RULES), respectively, to do playing techniques transfer. For example, if you want to transfer a piece of music that belongs to the Northern school style, into the style of the Southern school, you only need to apply the trained Southern school's playing techniques tagging models to the piece of music.

5.3 Reconstruction Result

After doing melody reconstruction and playing techniques reconstruction, the final reconstruction results can be gotten. A reconstruction example that shows the music style transfer result from the Southern school to the Northern school can be seen in Fig. 6. In each subfigure, the first row represents notes in the staff, the second row represents notes and playing techniques in the form of numbered musical notation. It can be seen that with the increase of iteration number, more kinds of playing techniques related to the Northern school (NT) is been generated. Adding the original melodies with playing techniques using our models, only tonguing appears two times. After 20 iterations of melodies, tonguing appears more times than the original melodies. After 60 iterations, typical playing techniques with the Northern school features like flutter tonguing, portamento appear, too.

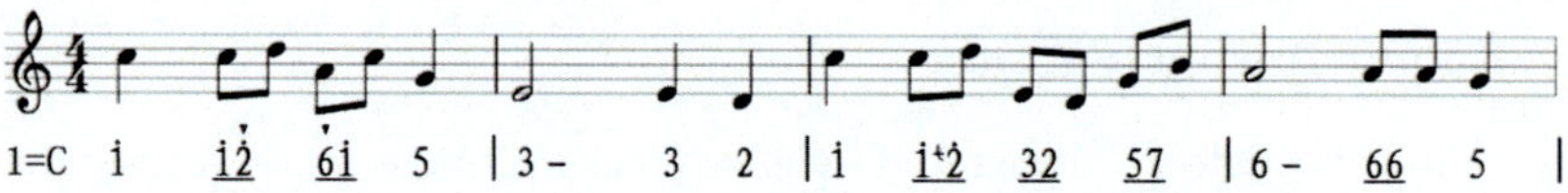

(a) Original melodies adding generated playing techniques

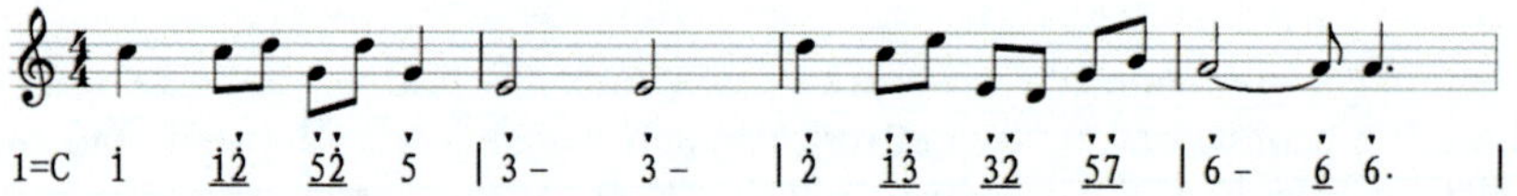

(b) Generated melodies after 20 iterations adding generated playing techniques

(c) Generated melodies after 60 iterations adding generated playing techniques

Fig. 6. A transfer example from the Southern school to the Northern school. The original melodies come from *Song of Soochow.*

5.4 Reconstruction Evaluation

After getting reconstruction results, we did the evaluation. In the reconstruction evaluation, we set four subtasks in total: the Northern school to the Southern school (N2S), the Southern school to the Northern school (S2N), the other school to the Northern school (O2N), the other school to the Southern school (O2S). After getting reconstruction results in the form of symbolic music, we played them in Dizi to get audios. We made a questionnaire to do an evaluation. There are 35 participants in total, and all of them have related music background. The score of evaluation is from 1 to 10. A higher score means a more thorough reconstruction. For the music style transfer task, it is not only needed to transfer to the target style, but also needed to maintain the original content, so it is a good result to get an upper-middle score.

The evaluation results are shown in Table 3. In general, we can see that for all four subtasks, the score of using both melody reconstruction and playing techniques reconstruction, is higher than the score of using only melody reconstruction. Besides, we can see that BILSTM-CRF performs better when the Northern school is seen as the music transfer target, while BILSTM-RULES performs better when the Southern school is seen as the transfer target.

Table 3. Results of audience evaluation. In this table, BILSTM-CRF and BILSTM-RULES are both methods of playing techniques reconstruction

Method	S2N	N2S	O2N	O2S
Only melody reconstruction	4.79	6.00	5.82	5.50
Melody reconstruction + BILSTM-CRF	5.41	6.36	6.79	6.29
Melody reconstruction + BILSTM-RULES	5.18	6.50	6.15	6.65

6 Conclusions and Future Work

In this paper, we proposed a general collection method of Chinese music and collected a Dizi symbolic dataset. Using some machine learning methods, we trained some classification models which laid the foundation for music reconstruction and found some interesting phenomena. We also gave a reconstruction example about music style transfer, where audience tests were done to do the evaluation.

Future work includes maintenance and expansion of the Dizi dataset, as well as broader and in-depth applications and research based on this dataset.

Acknowledgement. Supported by MOE (Ministry of Education in China) Youth Project of Humanities and Social Sciences, No. 19YJCZH084.

References

1. Cuthbert, M.S., Ariza, C.: Music21: a toolkit for computer-aided musicology and symbolic music data. In: 11th International Society for Music Information Retrieval (2010)
2. Dai, S., Xia, G.: Computational modeling for common pipa techniques. J. Fudan Univ. (Nat. Sci.) **57**(3), 2018–2371 (2018)
3. Graves, A., Jaitly, N., Mohamed, A.R.: Hybrid speech recognition with deep bidirectional LSTM. In: IEEE Workshop on Automatic Speech Recognition and Understanding. IEEE (2013)
4. Graves, A., Mohamed, A.R., Hinton, G.: Speech recognition with deep recurrent neural networks. In: 38th IEEE International Conference on Acoustics, Speech and Signal Processing. IEEE (2013)
5. Hearst, M.A., Dumais, S.T., Osuna, E., Platt, J., Scholkopf, B.: Support vector machines. IEEE Intell. Syst. Appl. **13**(4), 18–28 (1998)

6. Huang, Z., Xu, W., Yu, K.: Bidirectional LSTM-CRF models for sequence tagging. CoRR (2015)
7. Kim, Y.: Convolutional neural networks for sentence classification. In: Conference on Empirical Methods in Natural Language Processing (2014)
8. Lafferty, J.D., McCallum, A., Pereira, F.C.: Conditional random fields: probabilistic models for segmenting and labeling sequence data. In: 8th International Conference on Machine Learning (2001)
9. Li, R., Ding, Y., Li, W., Bi, M.: Automatic interpretation of Chinese traditional musical notation using conditional random field. In: International Symposium on Computer Music Modelling and Retrieval (2012)
10. Li, S., Wu, Y.: An introduction to a symbolic music dataset of Chinese Guqin pieces and its application example. J. Fudan Univ. (Nat. Sci.) **59**(3), 276–285 (2020)
11. Li, Z.: Anthology of Zhen Li's Dizi Music. People's Music Publishing House (2003)
12. Liang, X., Li, Z., Liu, J., Li, W., Zhu, J., Han, B.: Constructing a multimedia Chinese musical instrument database. In: 6th Conference on Sound and Music Technology (2019)
13. Luo, J., Yang, X., Ji, S., Li, J.: MG-VAE: deep Chinese folk songs generation with specific regional styles. In: 7th Conference on Sound and Music Technology. Springer (2020)
14. Mikolov, T., Chen, K., Corrado, G., Dean, J.: Efficient estimation of word representations in vector space. In: 1st International Conference on Learning Representations (2013)
15. Jones, K.S.: A Statistical Interpretation of Term Specificity and Its Application in Retrieval. Taylor Graham Publishing (1988)
16. Wang, C., Benetos, E., Chew, E., et al.: CBF-periDB: a Chinese bamboo flute dataset for periodic modulation analysis. In: 20th International Society for Music Information Retrieval Conference (2019)
17. Wang, C., Lostanlen, V., Benetos, E., Chew, E.: Playing technique recognition by joint time–frequency scattering. In: 45th IEEE International Conference on Acoustics, Speech, and Signal Processing, pp. 881–885. IEEE (2020)
18. Wang, H.: Research on Chinese traditional bamboo flute playing techniques. Master's thesis, Shaanxi Normal University (2014)
19. Xie, Y., Li, R.: Symbolic music playing techniques generation as a tagging problem. arXiv preprint arXiv:2008.03436 (2020)
20. Yan, N., Yu, Y.: Collection of famous Chinese bamboo flute music. Shanghai Music Publishing House (1994)
21. Yang, L., Chew, E., Rajab, K.Z.: Cross-cultural comparisons of expressivity in recorded erhu and violin music: performer vibrato styles. In: Workshop on Folk Music Analysis (2014)

Development of a Virtual Yangqin App with Unity Based on the Audio Object Pool Pattern

Ke Lyu[1] and Rongfeng Li[2](✉)

[1] Beijing University of Posts and Telecommunications, No 10, Xitucheng Road, Beijing, China

[2] Beijing Key Laboratory of Network System and Network Culture, Beijing University of Posts and Telecommunications, No 10, Xitucheng Road, Beijing, China

lirongfeng@bupt.edu.cn

Abstract. The yangqin, or the Chinese dulcimer, is a traditional Chinese plucked string instrument with a long history. This article designs and develops a virtual musical instrument App for the yangqin, presenting a three-dimensional instrument on the interface for users to perform with no difference from real instruments in audition and vision. The yangqin is a fixed-pitch instrument containing hundreds of strings with each pitch running in courses. The article starts the study with the design of the audio object pool pattern for realistic acoustic characteristics of sounding quickly repeated notes of fixed-pitch instruments. The following part focuses on the application development on both iOS and Android platforms with Unity. Finally, the article explores a new way of education for Chinese instruments and designs the teaching system for the App.

Keywords: Virtual musical instrument · Yangqin · Object pool · Unity · App

1 Introduction

Technology plays an important role in the world of professional audio production [1]. The prosperity of the Internet is providing a universal platform for music production, giving the birth of smart musical instruments on mobile devices. A variety of musical instrument apps are available, which permit users to play an iPad or an Android tablet as though it were a particular musical instrument and some scholars regard the iPad itself as a legitimate musical instrument [2]. However, Chinese musical instrument application Chinese traditional instruments are in the minority in the fields of computer music for historical reasons, whereas the popularization of mobile smart devices provides a new chance for the inheritance and development of these instruments. At present, application developments for Chinese instruments, though few, have been in a good trend. The extensive sound library of traditional Chinese instruments – the pipa, erhu, guzheng and Chinese percussion has been added in GarageBand for iOS since 2016 [3]. There are also Chinese software companies that develop creative Chinese

X. Shao et al. (Eds.): CSMT 2020, LNEE 761, pp. 29–42, 2021.
https://doi.org/10.1007/978-981-16-1649-5_3

instrument Apps with high-quality sound sources such as the Hulusi from Yinyueba [4] and the Qudi and Suona from JamKoo [5].

However, these products generally share the following deficiencies. First, in most Apps the instrument performance technique cannot be realized in an authentic way, and the sensitivity while playing the instruments is relatively poor so that smooth performances could not be achieved. Second, the interface is mostly unaesthetic and unable to show the complete three-dimensional models of the instrument. Third, most Apps include the education part, which is a great innovation, but only with some embedded video courses. It is inconvenient for amateur users to repeatedly switch the scenes to learn how to play the instruments provided in the Apps.

Most importantly, up to now there are no instrument applications for the yangqin either on computers or mobile platforms. Serving as the accompaniment in Chinese instruments orchestra, the yangqin plays the same important role as the piano in western music. However, the yangqin has not been promoted at schools at present, partly because the yangqin is too large in size and inconvenient to carry, and the investment of many schools in music education is limited for supporting the purchase of yangqin in large numbers [6]. A virtual yangqin App can be a perfect alternative on occasions where the volume of the real instrument cannot be sustained. As a fixed pitch instrument, each course of strings corresponds only one definite pitch, easy for digitization.

Based on the above, the article selects the yangqin as the target instrument for the design and development of the virtual instrument application on mobile platforms. The research work will be carried out from three aspects: audio synthesis, application development and teaching system design. The specific research tasks are as follows:

First, algorithm design of the audio object pool pattern for audio synthesis processing for fixed-pitch instruments. The algorithm is aimed at achieving tremolo and other quick playing or repeating effects in the virtual yangqin. The audio source of the yangqin comes from **the Multimedia Chinese Musical Instrument Database** [7] established in China Conservatory of Music.

Second, application development of the virtual instrument with Unity 3D for iOS and Android platform. The implementation of basic playing and other different functions, as well as the interaction design are included.

Third, program design of the teaching system, including follow-up learning mode and automatic performing mode for a Chinese classical music work. In follow-up learning mode, the program will highlight the strings to be played and check off the results. In automatic performing mode, the notes will be sounded automatically according to the correct rhythm and pitch of the music score.

2 The Audio Object Pool Pattern for Fixed-Pitch Instruments

Percussion instruments are classified into instruments with definite pitch and with indefinite pitch in **Handbook of Percussion Instruments** [8] by Karl Peinkofer. In this article, the concept of fixed-pitch instruments extends from percussion and refer to instruments of all kinds whose sound unit data and musical pitch data have a one-to-one mapping relationship, such as the marimba, yangqin and chimes.

According to the rule of instrument vibration, when a fixed-pitch instrument is played with different pitches, each note stays sounded and fades naturally to silence with no interference from others, while when repeated with the same pitch, the previous note will stop immediately at the moment when the next one sounds. In order to avoid the system sound delay and simulate the acoustic characteristics mentioned above in the best way, the article designs an algorithm for fixed-pitch instrument playing. The basic idea is: when playing with the same pitch, the second note is played first, and the first note fades to silence after a short transition time.

2.1 Audio Object Pool Management

In Unity, all the objects must be instantiated, and the audio source that controls the game sound is an instantiated object in the game scene. And an audio clip must be put into an audio source for playing and stopping management, and other effect controls. For the solution above, each pitch is saved in an audio clip, but as it involves playing, stopping and fading controls for repeating, notes in different terms must be loaded into different audio sources for management. That will be a problem since every time a note is played, an audio source is needed. Real-time generation and destruction can consume huge amount of memory. Especially when producing a tremolo, a distinctive technique for yangqin, meaning the same pitch is repeated intensively, it can bring an unbearable load to the program, calling for optimal management of these audio sources.

The article refers to the concept and process of the object pool pattern in game development to manage audio sources in the scene. The book **Pro Design Patterns in Swift** [9] by Adam Freeman decomposes the basic operation of the object pool pattern into four operations. The first operation is initialization for the objects to be managed. The second one is checkout, in which a component borrows the object needed from the pool. The third one is the component using the object to perform work. The fourth one is check-in, where the component returns the object to the pool.

The audio object pool pattern designed in this article carry on the idea of the object pool, which is to activate and deactivate the instantiated objects in a fixed memory space. The audio sources outside the pool are in the status of activated and ready for the playing control program, while others inside the pool are deactivated and waiting in line. The system applies for an audio pool with a size of 10 for each sounding component, which contains 10 audio sources, all of which are loaded with the same audio file where the corresponding tone is written. Figure 1 shows the process of the audio pool. Every time a playing instruction is received, 2 of the 10 Audio Sources will be called out of the pool, and they will line up in a circular queue for that 2 called-out positions marked yellow in Fig. 1. The two called-out audio sources (audio A and audio B in Fig. 1) out of the pool participate in the controlling program in order.

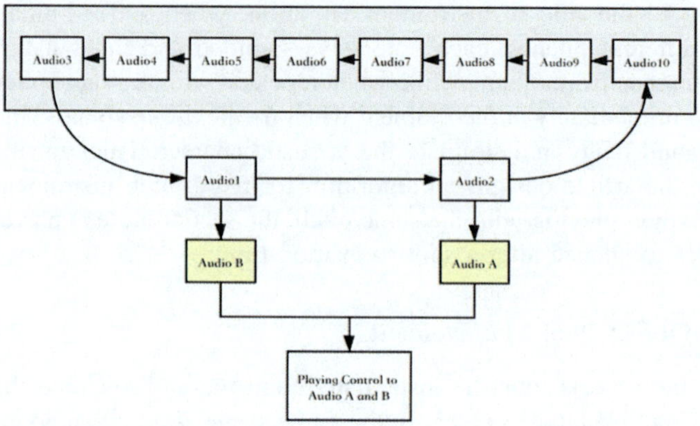

Fig. 1. Process of the audio management in the pool

2.2 Algorithm for Audio Playing in the Pool

For calling out and putting back the audio sources in the pool, a variable t is used to record the number of times the tone is played in the program, and a function `Get-ClipTerm (int)` of type int is declared, whose return value is `t – t /10 * 10` (the division '*t /10*' gives a round-off value), the remainder obtained by dividing *t* by *10.* The value is synchronized with the looping pace of the audio sources in the pool and is used for matching the sequence number of the audio sources in the pool 10 times as a round, and selecting 2 audio sources each time to run the controlling program. The specific steps are as follows:

1. At the beginning of the program, apply for 10 Audio Sources, compile them into the array `audioSource[]`, and load the audio clips corresponding to the pitch one by one.
2. Initialize the variable `t = 0`.
3. When the program is running, if the playing instruction is recognized.
   a. Play `audioSource[GetClipTerm(t)]`, which is audio B in Fig. 1.
   b. When *t* is not equal to *0*, judge whether `audioSource[GetClipTerm(t-1)]` (audio A) is playing. If playing, the volume will gradually decrease to 0 within 0.5s. If not, skip.
   c. When t equals 0, the note has not been played before. Skip it.
   d. Execute `t = t + 1`, which means the number of times the tone is played increases once.

3 Application Development with Unity

3.1 System Design

In the development of virtual yangqin App, Unity3D is used as the game engine for application development, and C# as the programming language, iOS and Android as target platforms released under the Mac OS X system. The App includes the following features:

1. can be played on each course of strings of the complete three-dimensional yangqin;
2. can switch the reverb effect, including concert hall, padded cell, forest, and cave;
3. can be switched on each string and display fixed name and different tone under the name;
4. involves the teaching system with follow-up learning mode and automatic performing mode for the traditional Chinese song Chun Dao Qing Jiang.

The system is functionally divided into 5 modules. The overall functional architecture is shown in Fig. 2.

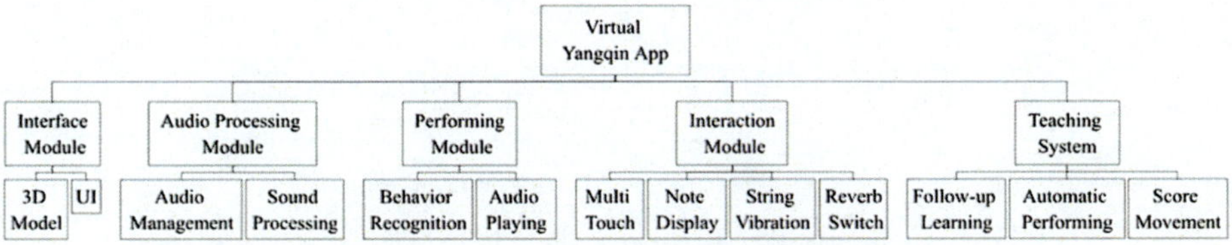

Fig. 2. System function architecture

The interface module is mainly composed of a three-dimensional instrument model and a UI. Some operations are performed directly on the model, and other functions need to be realized through the UI buttons. The audio processing module is divided into audio management and sound processing with the audio object pool pattern described above. The performing module is to realize the touch recognition and play the correct tone, including clicking the string to play through the collision body recognition, and then playing the audio-processed tones. The interaction module includes multi-touch, note display, string vibration and reverb switch.

The teaching system is based on the complete function of basic performance and will be specifically introduced in the next section. It provides the function of follow-up learning by program constraints as well as the display of real-time moving music scores by numbered musical notation. At the same time, the automatic performing mode is set for users to learn from, and the interface will also display the scores for users to read while playing. Buttons are set for users to switch among the treble part (playing with the left hand), the bass part (with the right hand) or both parts (with both hands), and to replay and exit.

3.2 Function Module Design and Implementation

Interface Module. The main part of the application is the instrument itself. A three-dimensional musical instrument model is on the basis of the 402 yangqin in shape, with a background image that matches the colors. The App interface is shown in Fig. 3. The structure of the UI system is shown in Fig. 4.

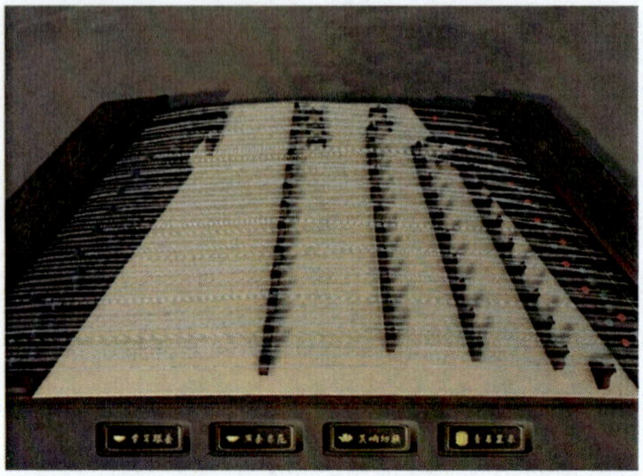

Fig. 3. The app interface

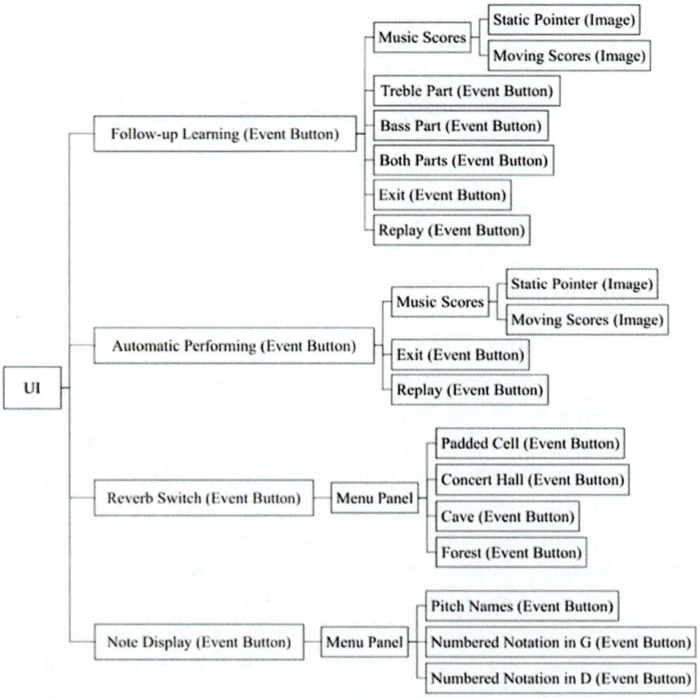

Fig. 4. The structure of the UI system

According to the hierarchical layer above, the four parent event buttons always appear below the interface, and secondary buttons and UI elements they contain (such as all buttons in the menu panel of the 'note display' button, and the music score in the teaching system) will not appear until their parent button is clicked. Figure 5 shows the activated menu panel of 'note display' giving three choices.

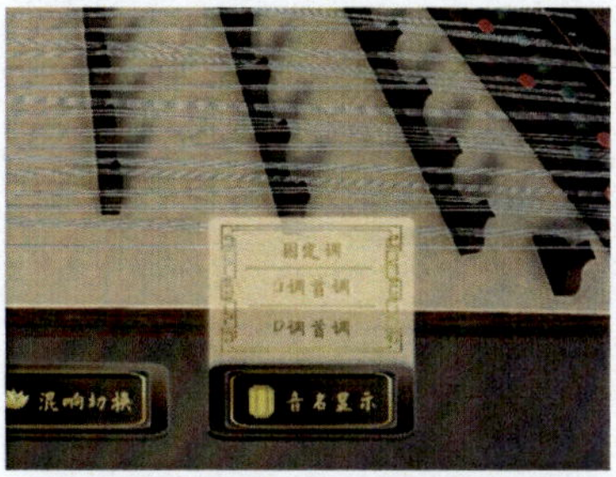

Fig. 5. The menu panel of note display

There are some mutually exclusive relationships in these buttons. For example, in the follow-up learning mode, when clicking on the button to choose which part to learn, the clicked one turns yellow indicating that the status is activated, as shown in Fig. 6, and the other two buttons, whether activated or not, will be inactivated.

Fig. 6. The 'both parts' button that is activated

Audio Processing Module. The audio object pool pattern introduced in Sect. 2 is used in the audio processing module. According to the acoustic characteristics of the yangqin, each note that is played stays sounded and fades to silence unless another is played again at the same pitch, and the previous note stops at once.

There are 61 courses of strings and 52 different pitches in a 402 yangqin. The audio files of 52 naturally decayed notes from F to a3, sampled and recorded from the real yangqin in the database are used for development. The files in different pitches are stored as audio clips in the game scene, each of which is given one audio object pool with 10 audio sources loading the pitch once the application is started. The operations in every pool such as checkout and check-in from the pool in the circular queue of audio sources and the playing controlling program work independently.

The audio processing script is attached to each game object of a set of strings. 10 audio sources are created at system startup time, and organized into an array `audioSource[]` to create an audio pool of 10. The audio clips at the relevant pitches are loaded then in sequence by matching the names with the game objects. The playing controlling part is written into the function void `stringPlaying()`, getting the audio source to play and stop when receiving the pitch repeating command. The function `GetClipTerm(int)` returns the remainder obtained by dividing t by 10, used to mark the position of audio sources in the cyclical audio pool. The audio source that are to be played instantly. The audio source `audioSource[GetClipTerm(t)]` is the one to be played instantly, and `audioSource [GetClipTerm(t-1)]` is the previous audio source. The flowchart of this module is shown in Fig. 7.

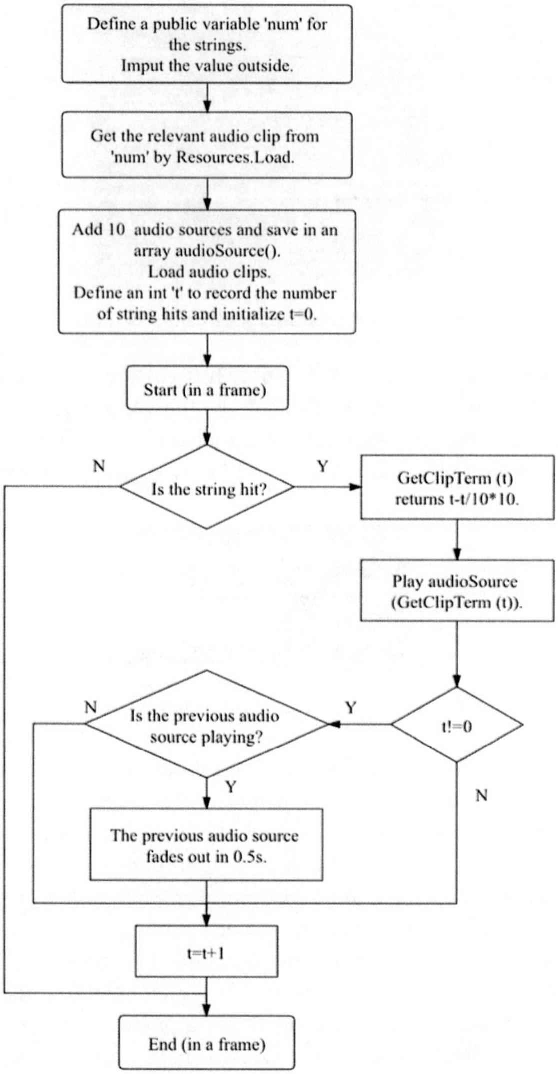

Fig. 7. Flowchart of the audio processing module

Performing Module. As a string vibration sounding system, a yangqin string sounds at a certain pitch depending on its thickness, tension, and effective vibration length. In theory, anywhere of a string can be hit, but to avoid the interference of other tangled strings nearby, there are conventional hitting positions when playing the yangqin. The virtual yangqin App follows this performing rule and sets the effective performing part as is framed in Fig. 8. In the game scene in Unity, 3D box colliders are added on the strings within these performing positions. Only when the ray emitted from the screen is detected to collide with a string can the string vibrate and sound.

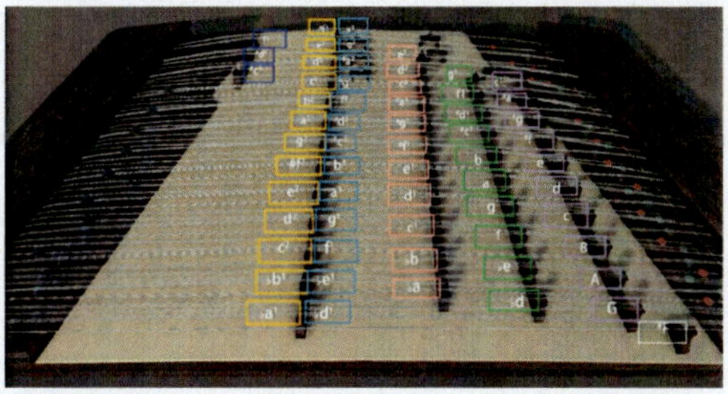

Fig. 8. The effective performing part in the virtual Yangqin

The authors have designed and made a pair of capacitive hammers as accessories for the virtual yangqin App to simulate the effect of real yangqin performance to the greatest extent. As shown in Fig. 9, the bamboo hammers are wrapped up in tinfoil, and the tips of the capacitive pens made of conductive silica gel are sleeved on the top as a contact point. Silicone conducts human body capacitance to the mobile touch screen to achieve a touch effect. Figure 10 shows the application scenario. The capacitive hammers retain the elasticity of the bamboo, and are able to bounce on the screen and achieve quick combos.

Fig. 9. The capacitive hammers

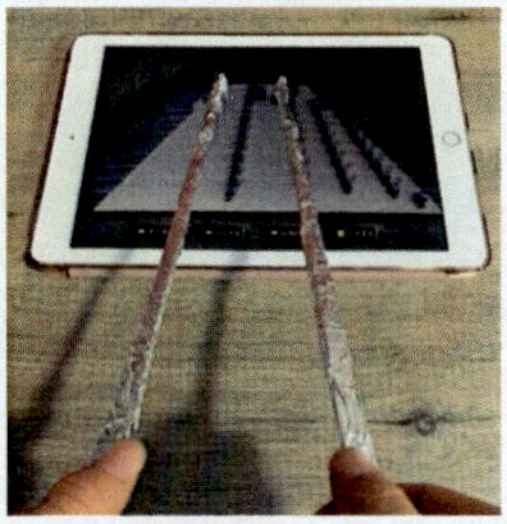

Fig. 10. Performing on the virtual Yangqin with the capacitive hammers

Interaction Module. The interaction module mainly includes the following parts:

1. Multi-touch. The Yangqin allows one or two strings to be hit to sound. More than two fingers will not be recognized.
2. Note display. The yangqin is with fixed pitch, similar to the piano, but its arrangement is not like the piano from low to high with a fixed pitch rising. Instead, it is with less

regularity. Most folk music is performed in movable-do system, so for helping amateur users to play the yangqin without remembering the arrangement of the pitches, the pitch names as well as the numbered notation in G and D are provided for display. Figure 7 above shows the interface with the pitch names showing on.
3. Strings vibration. When the strings are hit, the strings exhibit a vibration effect.
4. Reverb switch. There are four reverb effects: padded cell, concert hall, cave, and forest.

4 Design and Practice of Teaching System

4.1 Teaching Mode Design

It is required for piano learners to learn sheet music notation and its mapping to respective piano keys, together with articulation details [10]. Taking the portable piano tutoring system Mr. Piano developed by Sun and Chiang [11] as a reference, the teaching system of the virtual yangqin designed in this article involves two modes: follow-up learning and automatic performing. The preset learning repertoire is 'Chun Dao Qing Jiang'. In the follow-up learning mode, the user plays notes beat by beat with the instruction on the screen. Only when each beat is played correctly will the system continue. In the automatic performing mode, the system plays the track automatically.

Follow-up Learning Mode. As shown in Fig. 11, in the follow-up learning mode, the score of 'Chun Dao Qing Jiang' will be displayed at the top. The user needs to click on one of the three buttons on the left side of the score to start learning, where 'treble part' means to play the upward single part of the score with the left hand, 'bass part' refers to the downward part with the right hand, and 'both part' refers to the complete music with both hands.

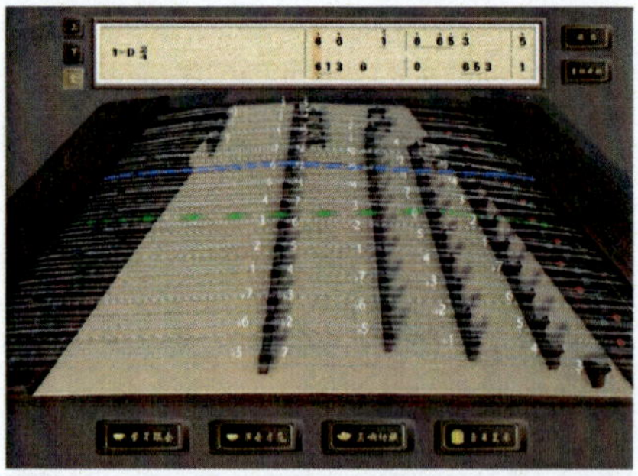

Fig. 11. The follow-up learning mode

After selecting the part, the string corresponding to the pitch to be played in the first beat will be highlighted, with blue indicating the treble part and green indicating the bass part. For example, in Fig. 11, the 'both parts' button is clicked on. The system will maintain this state until all the notes of this beat are recognized correctly. After the correct performance is recognized, it will move on to the next beat, and the score will move forward in the preset song speed. When the pointer overlaps the second note, the system will stop again, and the strings to be played in the second beat will be highlighted. If it is recognized that the user has played the wrong string, the string will flash slightly in red.

Click the 'restart' button and it will go back to the beginning of the track and relearn, and the highlighted part will be cleared. Click the 'Exit' button and the follow-up learning mode will be exited.

Automatic Performing Mode. In the automatic performing mode, the system automatically plays according to the content of the score. The system's built-in audio source at a certain pitch is called to directly play the corresponding note in accordance with the specified rhythm. In the automatic performing mode, the score will still be displayed on the top, and the score will move synchronously with the performance throughout the entire process, helping users to better understand the melody and rhythm. When the system automatically sounds, the user can still click on any course of strings on the yangqin and play the notes for practice or accompaniment.

Similarly, click the 'Restart' button and the track will be replayed, and the highlighted part will be cleared. Click the 'Exit' button and the automatic performing mode will be exited.

4.2 Implementation of the Teaching System

Music Score Compilation. In this system, the method of compiling scores is to use a certain note length as a rhythm unit (the 6th note is used in this article), and compile each note into an array of string type, and the content is its pitch name. The rests are represented by null. The rest here is not just a pause in musical theory, but is stretched to the circumstances that no musical note is to play at this 16th note moment. The treble and bass parts are compiled separately. The following is the compiled score of the first sentence of treble part of 'Chun Dao Qing Jiang', stored in a string type array.

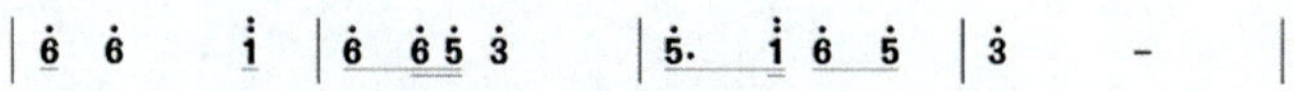

Fig. 12. The first sentence of treble part of 'Chun Dao Qing Jiang' in numbered musical notation (in the key of D)

Figure 12 gives the corresponding content in the original numbered musical notation.

```
"b2",null,"b2",null,null,null,"d3",null,
"b2",null,"b2","a2","#f2",null,null,null,
"a2",null,null,"d3","b2",null,"a2",null,
"#f2",null,null,null,null,null,null,null,
```

Follow-up Learning Program. The follow-up learning mode consists of three choices for the playing part: the treble part, bass part and both parts. The same program is used for the treble and bass part learning. The basic idea is to use the integer variable beat to record the number of beats and control the loop. The program flow chart is shown in Fig. 13.

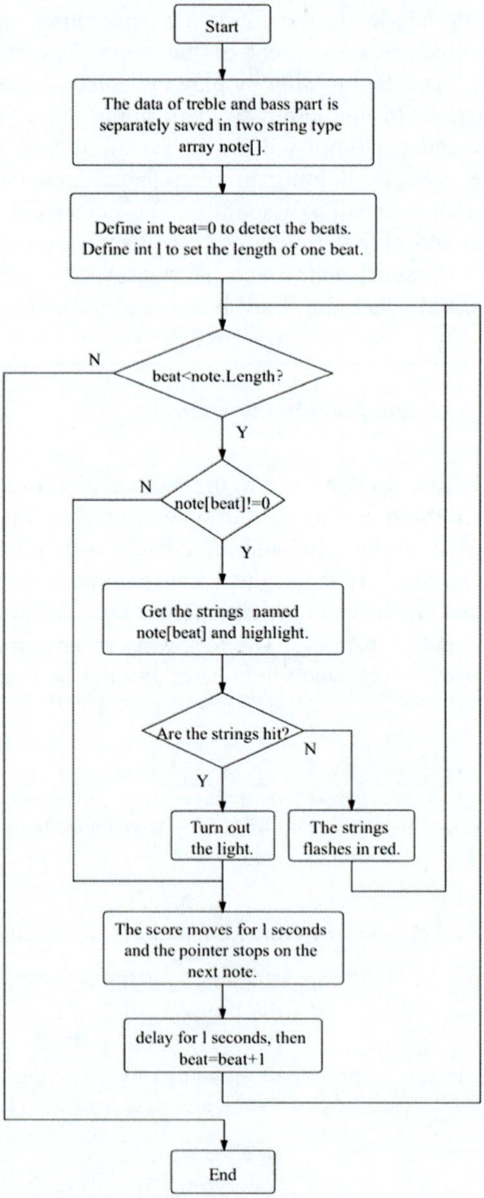

Fig. 13. Flow chart of the follow-up learning program for a single part

The program for the both parts involves the issue of multi-touch. For a duet for two pieces of melody, every beat should be distinguished from single-tone or double-tone. In the case of double-tone, the two-finger touch and single-finger touch situations will also be considered. It is because the conditions for the system to judge the two-finger touch are much stricter than the actual double-tone recognized by human ears. It is difficult for users to achieve standard two-finger touch, so single-finger performing must be allowed. In this program, for single-finger double-tone performing, only when the two target notes are recognized within one beat's length will the performing be considered correct. If a wrong note occurs, it will be judged wrong even though one or two correct notes have already been recognized in that beat.

5 Conclusion

The development of Chinese musical instruments is still sluggish in the current world of computer music, while Chinese traditional music is a valuable historic asset for the world and requires protection and promotion. The authors hope that the design and development of a complicated and realistic virtual App of a Chinese instrument can take a small step.

Authenticity is the biggest feature and advantage of the research results of the article.

First, the timbre is real. The audio used in this article is sampled and synthesized based on real timbre, and the audio object pool pattern is designed to simulate the acoustic characteristics and ensure that the virtual yangqin sounds the same as the real one when playing.

Second, the appearance is realistic. The shape of the virtual yangqin model is completely following the structure of the real instrument. Other features on simulating its appearance include string vibration and the effective performing positions.

Third, the performing method is authentic. The App completely refers to the playing method of the yangqin, and aims to help all users to learn and play in a correct way from the teaching system.

It is hoped that based on the high-quality and large-scale Chinese music database, more Chinese musical instrument applications can be further realized, so that Chinese musical instruments can make greater progress on the road of intelligence and generalization.

Acknowledgement. Supported by MOE (Ministry of Education in China) Youth Project of Humanities and Social Sciences, No. 19YJCZH084.

References

1. Barro, S.J., Fernández-Caramés, T.M., Escudero, C.J.: Enabling collaborative musical activities through wireless sensor networks. Int. J. Distrib. Sensor Netw. **8**(3), 314078 (2012)
2. Kang, S.: The Effect of Motivation on Students' Preference for Acoustic or iPad Instruments: Comparing Guitars and Gayageums. University of Florida, Gainesville (2016)

3. GarageBand for iOS. https://www.apple.com/ios/garageband. Accessed 2 Aug 2020
4. iOS Application. Yinyueba. https://apps.apple.com/cn/app/yin-yue-ba/id989562871. Accessed 2 Aug 2020
5. iOS Applications. JamKoo-Live Performance Synth. https://apps.apple.com/app/jamkoo/id1457240484. Accessed 2 Aug 2020
6. Li, X.: Analysis on the current situation of yangqin's application in music education in primary and middle schools. Home Drama (10), 173 (2020)
7. Liang, X., Li, Z., Liu, J., et al.: Constructing a multimedia Chinese musical instrument database. In: Wei, L., Shengchen, L., Xi, S., Zijin, L. (eds.) Proceedings of the 6th Conference on Sound and Music Technology (CSMT), LNEE, vol. 568, pp. 53–60. Springer, Singapore (2019)
8. Peinkofer, K., Tannigel, F.: Handbook of Percussion Instruments. Schott, London (1976)
9. Freeman, A.: The object pool pattern. In: Pro Design Patterns in Swift, pp. 137–155. Apress, Berkeley (2015)
10. Rogers, K., Röhlig, A., Weing, M., et al.: P.I.A.N.O.: faster piano learning with interactive projection. In: Proceedings of the Ninth ACM International Conference on Interactive Tabletops and Surfaces, pp. 149–158. Association for Computing Machinery, New York (2014)
11. Sun, C., Chiang, P.: Mr. Piano: a portable piano tutoring system. In: 2018 IEEE XXV International Conference on Electronics, Electrical Engineering and Computing (INTERCON), pp. 1–4. IEEE, Lima (2018)

Channel-wise Attention Mechanism in Convolutional Neural Networks for Music Emotion Recognition

Xi Chen[1], Lei Wang[1], Andi Pan[1], and Wei Li[1,2(✉)]

[1] School of Computer Science, Fudan University, Shanghai 201203, China
{19210240230,18210240192,18210240151,weili-fudan}@fudan.edu.cn
[2] Shanghai Key Laboratory of Intelligent Information Processing, Fudan University, Shanghai 201203, China

Abstract. Music Emotion Recognition (MER), a subject of affective computing, aims to identify the emotion of a musical track. With the fast development of deep learning, neural networks, such as Convolutional Neural Network (CNN) and Long Short-Term Memory (LSTM), have been applied recently. However, while using convolutional kernels, channels are treated equally, which means treating different aspects (such as tempo and vibrato related features) of a music clip equally. It's against the rule of human perception. Therefore, Channel-wise Attention Mechanism is introduced into the task of Music Emotion Recognition. The performance could be improved to a certain extent.

Keywords: Music Emotion Recognition · Channel-wise Attention Mechanism

1 Introduction

Music Emotion Recognition (MER), a subject of both Music Information Retrieval and Affective Computing, aims to identify the emotion conveyed by a musical clip [18]. Driven by large demands in the music industry such as providing a content-oriented categorization scheme, generating playlist automatically and music recommendation [6,10,22], MER has developed rapidly in recent years [1].

Traditional methods are feature-based ones. The most commonly used acoustic features (e.g. Mel Frequency Cepstrum Coefficient, spectral shape in timber, Chromagram and Rhythm strength) are summarized in [9]. On the one hand, since there are hundreds and thousands of features to be considered, feature selection methods [16] for removing redundant ones and principle component analysis (PCA) methods [14] for dimension reduction are introduced. On the other hand, manual features sophisticated designed to express the nature of music emotion have always been an interest in the field. Since most features are low level and related to tone color, in 2018, [15] designed musical texture related and expressive technique related features.

X. Shao et al. (Eds.): CSMT 2020, LNEE 761, pp. 43–54, 2021.
https://doi.org/10.1007/978-981-16-1649-5_4

In recent years, with the development of computer hardware and the large available online data, deep learning has demonstrated its great power in many fields including Music Emotion Recognition. Instead of designing specific features which is a challenging task [12] and demands large human labor, a Neural Network itself could extract the very pertinent representations.

Besides simply using deep learning models, such as CNN used in [12], different mechanisms are used to improve the performance. Similar to the multitask learning theory, hoping the first several layers to extract commonly acoustic features and the last several layers to extract target-oriented features, [13] stacked one CNN layer with two RNN branches for arousal and valence regression. Some hope to utilize auxiliary information. Inspired by speech emotion recognition tasks considering spoken content [19], [5] proposed a multimodal architecture based on audio and lyric. In [11], additional harmonic and percussive features are fed into the bi-directional LSTM model. Because of the lacking of training data, others also use the transfer learning method, aiming to make use of excellent features in related domains [3].

As we can see above, despite different mechanisms, the base architectures are usually CNN. It is known that a convolutional neural network learns hierarchical features from level to level [21] and that a higher-layer feature maps depend on lower-layer maps [2]. For instance, in the early layers, low-level information such as tempo, pitch, (local) harmony or envelop might be extracted [3], while high-level semantic patterns such as expressivity and musical texture features would be detected [15] in later layers. However, on the one hand, while doing convolutional operations, the low-resolution features which contain abundant low-frequency information are treated equally across channels [23] (i.e. tempo and pitch information may not contribute exactly the same), hence, the extracted features are not powerful enough. On the other hand, without processing the whole music clip, only understanding a few important aspects, one could recognize its emotion, whereas, a CNN would process all the feature maps which is in contrast to human perception [4].

Fortunately, the problems existed could just be solved by the Channel-wise Attention Mechanism, which has been successfully applied in Computer Vision [20], Natural Language Processing, and Speech Processing. The Channel-wise Attention Mechanism could re-weight feature maps in channels. Moreover, since a feature map is computed from earlier ones, it is natural to apply attention mechanism in multiple layers [2]. In this way, multiple semantic abstractions could be gained [2]. The applied Channel-wise attention mechanism is a sophisticatedly chose one, detailed in Sect. 2.

Music Emotion Recognition tasks could be categorized into a classification one and a regression one. The proposed method is tested on both tasks and the performance has been improved. It should be mentioned that, since public music emotion classification datasets are small, which will even limit the performance of the baseline network, a larger music emotion classification dataset is made.

In summary, the contribution of this paper could be put as follows:

(I). To solve the problem of treating each feature map equally while recognizing musical emotion patterns, Channel-wise Attention Mechanism is applied in multi-layers.
(II). The Channel-wise Attention Mechanism is a sophisticated chosen one.
(III). The procedure of how to make a large musical emotion classification dataset is introduced.
(IV). The proposed method could be proven useful to a certain extent.

2 Proposed Method

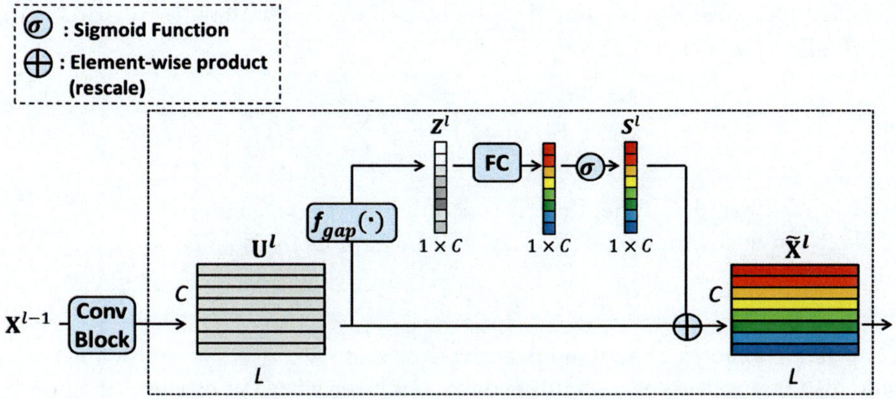

Fig. 1. The Channel-wise Attention Mechanism for Music Emotion Recognition

There are many sophisticated channel-wise attention mechanisms in literature, such as that in SENet [7], that in RCAB [23]. Obviously, we should not simply draw one of them to use, we should choose or design one on our needs. Firstly, to fully consider the interrelationships among all channels, the channel-wise attention mechanism is designed with the fully connected layer and the activation function, like that in SENet [7], rather than using convolutional ones whose receptive field is limited to only a few channels, such as that in RCAB [23]. Secondly, to learn a non-mutually-exclusive relationship, some channel-wise mechanism is under the mode of an encoder and a decoder scheme. However, after an encoder operation, whether it depends on doting with a weight matrix or convolutional operations, the rank is decreased after encoding, meaning some information from the feature map (though less important) will be lost. This is undesired. Thirdly, considering of computational efficiency, the designed channel-wise attention mechanism is lightweight.

Next, the proposed channel-wise attention mechanism and the backbone architecture will be introduced.

2.1 Channel-wise Attention Mechanism

The channel-wise attention mechanism block is a transformation block. On the above reasons, it is designed with a simple gating mechanism with an activation function. Figure 1 illustrates the mechanism along with the operations and the variables.

As a whole, it is a reweighting operation from $\mathbf{U}^l \in \mathbb{R}^{L \times C}$ to $\tilde{\mathbf{X}}^l \in \mathbb{R}^{L \times C}$, where $\mathbf{U}^l$ is the output feature map with the length of L and channel number of C after the l-th convolutional block with input $\mathbf{X}^{l-1}$. All of the superscripts in notation refer to the layer index.

First, each convolutional kernel with a fixed size receptive field serves as a local semantic information extractor. Therefore, each or some of the value in a feature map could not represent what the channel learns [7]. To mitigate this problem, the channel-wise statistic $\mathbf{Z}^l = \left[{z_1}^l, {z_2}^l, \ldots, {z_c}^l, \ldots, {z_C}^l\right]$ obtained by using global average pooling $\boldsymbol{f}_{gap}(\cdot)$ is used as the channel feature descriptor. For detail, ${z_c}^l$ is calculated by:

$$z_c^l = \boldsymbol{f}_{gap}\left(\boldsymbol{x_c}^l\right) = \frac{1}{L}\sum_{i=1}^{L} \boldsymbol{x_c}^l(i) \tag{1}$$

More sophisticated channel descriptors could also be considered.

Next, inter-channel dependencies will be exploit by Eq. (2):

$$\mathbf{S}^l = \sigma\left(g\left(\mathbf{Z}^l\right)\right) = \sigma\left(\mathbf{W}^l \cdot \mathbf{Z}^l\right), \tag{2}$$

Where $\sigma(\cdot)$ denotes the sigmoid activation and $\mathbf{W}^l \in \mathbb{R}^{C \times C}$. Obviously, $g(\cdot)$ could also be interpreted as a fully connected layer with $\mathbf{W}^l$ as the corresponding parameter.

Finally, the attended feature map could be obtained by modulating $\mathbf{U}^l$ with $\mathbf{S}^l$, for each channel

$$\tilde{x}_c^l = f_{\mathbf{rs}}\left(u_c^l, s_c^l\right). \tag{3}$$

In Eq. (3) $f_{\mathbf{rs}}$ means rescaling u_c^l with scalar s_c^l.

2.2 Backbone Architecture

The backbone architecture we adopted is following the audio subnet of the Audio-Lyric Bimodal in [5]. It is originally used for the emotion value regression task. While in this paper, the backbone architecture would be applied to both regression and classification with different outputs.

It is composed of two convolutional blocks and two dense blocks. For one thing, as for the convolutional block, a convolutional layer, a max pooling layer and batch normalization are consecutive. The (the number of kernels, kernel size, stride) for convolutional layer are (32, 8, 1) and (16, 8, 1) separately, while the (kernel size, stride) for the max pooling layer are all in (4, 4). For another thing, the dense block includes a dropout and a fully connected layer. The intermedia neural number for the two dense blocks is 64 [5].

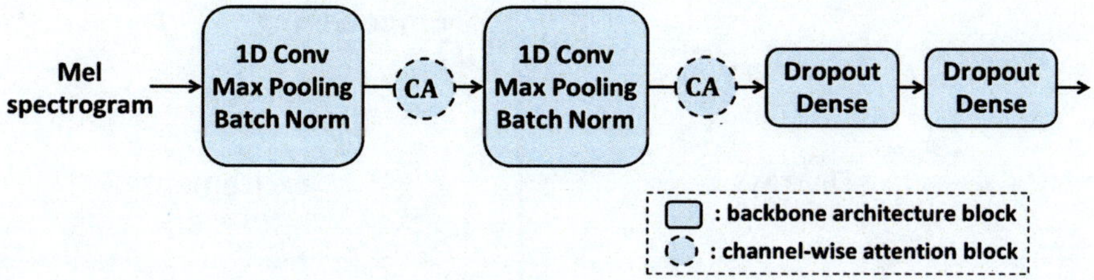

Fig. 2. Music Emotion Classification Model with dotted line representing the attention block and solid line representing the backbone architecture block

3 Evaluation

Datasets, metrics and experiment settings will be talked here. More, the method for making a large music emotion classification dataset is presented under the hope of helping other researchers to design much more powerful systems.

3.1 Dataset

Music Emotion Recognition tasks tend to use either categorical psychometrics or scalar/dimensional psychometrics for classification or regression [9]. Both music emotion representations are under the supporting of psychological theories [9]. Under categorical approaches, emotion tags are clustered into several classes. The well-known MIREX Audio Mood Classification Competition just uses this kind of psychometric [8]. While under continuous descriptors, a certain kind of emotion could be represented by a point in the Valence-Arousal (V-A) space [17]. Though there are two kinds of music descriptors, under the Circumflex Model of Affect [17], they could be transformed to each other.

Classification Task Dataset. For the reason that public music emotion classification datasets are small, containing only less than 1,000 clips, even the baseline deep learning neural network could not demonstrated its great power, not to mention the proposed channel-wise attention mechanism. Hence, under the guidance of the Circumflex Model of Affect by Russell [17], with the help of emotion related playlist (those which have been created intentionally and listened millions of times) on the mainstream music software, a large reliable dataset with thousands of music clips could be made with less human labor.

To begin with, a set of music emotion tags are chose according to human experience and psychology theory. Six tags, Stirring, Empowering, Angry, Somber, Peaceful and Upbeat are finally determined. Definitely, they are under the constrain of Circumflex Model of Affect, sparsely located on the model, which means that the gap between music emotion tags are large enough to make them separate well. Figure 3 will illustrate the relationship between the music tags and the Circumflex Model of Affect.

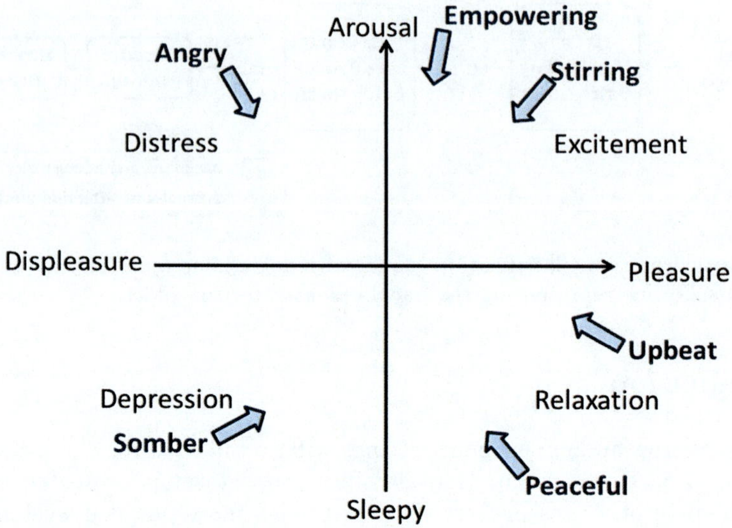

Fig. 3. Mapping the selected music emotion tags on the Circumflex Model of Affect

Next, searching the tag related playlists on popular music website such as NetEase cloud music and QQMusic, top played ones would be considered. By referencing to the comments of the playlists and humanly verifying each song, the final ones would be determined. After that, for each song, the first 5 s would be thrown away considering the emotion there might be different from the whole song, and then they would be cut into 30 s ones.

Finally, using this method, more than 4,000 music clips with sample rate of 44,100 are got.

Since annotated music excerpts are collected from website and are copyrighted, the dataset could not be made public. However, using the above mentioned method, researchers could make their own dataset easily.

Regression Task Dataset. As for the baseline architecture, the used continuous descriptor is song leveled, it uses an arousal value and a valence value to describe a 30 s music expert. Unfortunately, the dataset is not a public one. Mainstream public ones are all dynamically annotated, which means that they consider emotion variation in a music [1] and are annotated every once in a while. To mitigate this problem, we choose the averaged value of all annotations in a song to represent the song level descriptor.

For the dynamically annotated public dataset, we utilize a largest one, Emotional Analysis in Music (DEAM) [1]. It contains 1,802 songs including 1,744 45 s clips and 58 full length clips. The time resolution for annotation 2 Hz, meaning annotating per 500 ms. The annotated values are scaled in $[-1, +1]$.

In our experiment, since baseline architecture is designed for 30 s clips, only the middle 30 s (from the 7th to the 36th second of the clip) audio is preserved.

And, 58 full length clips are too long, using the song level averaged annotation to represent each segment is not a smart choice, therefore, they are thrown away. Finally, after processing, 1,744 clips are remained.

3.2 Metric

In notation, N, y^i, $\hat{y^i}$ corresponds to the number of samples, predicted label/value and real label/value separately, where $i \in [0, N-1]$.

Metric for Music Emotion Classification Task. Accuracy score, shorten as acc, represents the ratio of correctly classified samples to total number, could be written as:

$$acc\left(\hat{y^i}, y^i\right) = \frac{1}{N}\sum_{i=0}^{N-1} 1\left(\hat{y^i} == y^i\right). \tag{4}$$

Confusion matrix shows more detail information than acc. It is much easier to see how the system confusing among them. Let $\mathbf{C}$ to be matrix, $\mathbf{C}_{i,j}$ means the proportion of sample observed in class i but classified into class j. Specifically, $\mathbf{C}_{i,i}$ corresponds to the accuracy score for the i-th class.

Matric for Music Emotion Regression Task. Root Mean Square Error (RMSE) is a typical matric for regression tasks, meaning how far is the predicted value from the real one.

$$RMSE = \sqrt{\frac{1}{N}\sum_{i=1}^{N-1}\left(\hat{y^i} - y^i\right)^2}. \tag{5}$$

3.3 Settings

As mentioned in Sect. 2, the backbone architecture followed from [5] will be used for both regression and classification tasks.

For each audio clip, after upsampling to 44,100 Hz, an Mel-spectrogram is extracted with 40 mel bands, 1024 sample long Hann window with no overlapping as input [5]. In baseline, it uses pitch shifting to argument the data. However, pitch is an emotion related feature [11,15,18]. Therefore, data argument method in baseline is not adapted.

While training, we use Cross Entrpy loss for classification task, Mean Square Error loss for regression task, Adam is the optimizer.

4 Experiments and Results

In this section, experiments conducted to validate the effectiveness of the proposed method will be presented. In notation, AudioNet represents the baseline, Layer1, Layer2, LayerALL means the location of the added channel-wise attention mechanism (i.e., Layer1 meaning adding the attention mechanism after the first layer).

4.1 Validating the Proposed Method

In the first set of experiments, we would like to validate the power of the proposed attention mechanism's power and that of the multi-layer attention scheme by using the classification dataset. Since the baseline has two convolutional blocks, three architectures'performances will be evaluated, AudioNet, AudioNet_Layer1, AudioNet_Layer2, AudioNet_LayerALL.

Table 1. Overall and class level accuracy score for the baseline and attention added ones in different locations.

Overall accuracy	
Network	On Whole Dataset
AudioNet	0.665
AudioNet_Layer1	0.695
AudioNet_Layer2	0.736
AudioNet_LayerALL	**0.741**

Class Level accuracy						
Network	Angry	Somber	Upbeat	Peaceful	Empower	Stirring
AudioNet	0.792	0.583	0.736	0.626	0.657	0.458
AudioNet_Layer1	**0.877**	**0.613**	0.725	0.712	0.614	0.615
AudioNet_Layer2	0.836	0.603	0.750	0.720	**0.754**	**0.789**
AudioNet_LayerALL	0.866	0.557	**0.895**	**0.818**	0.732	0.567

Experiment results will be seen in Table 1. As we can see, whether adding channel-wise attention after layer 1 or layer 2, the performance could be improved distinctly; this can verify channel-wise attention's ability. When adding attention mechanism after all layers, the performance could be improved further, this could demonstrate the rationality of adding attention to multi layers.

It is interesting to find that the architecture adding attention to the later layer will demonstrate more power than that adding to the earlier one. The reason behind might be that earlier layers extract low-level represents while later ones extract class-specific features [7]. Emphasizing more on class-specific features will help more than extracting better common music characteristics, for example, better musical texture features will help more than pitch features in music emotion recognition [15].

4.2 Performance on Classification Task

Performance in overall between baseline and the proposed one has been demonstrate above, Table 2 gives more detail by using confusion matrix.

Table 2. Confusion matrix for baseline and the proposed

AudioNet						
	Angry	Somber	Upbeat	Peaceful	Empowering	Stirring
Angry	**0.792**	0.028	0.028	0.000	0.083	0.069
Somber	0.000	**0.583**	0.042	0.188	0.021	0.167
Upbeat	0.014	0.028	**0.736**	0.000	0.167	0.056
Peaceful	0.009	0.261	0.017	**0.626**	0.000	0.087
Empowering	0.171	0.029	0.071	0.014	**0.657**	0.057
Stirring	0.125	0.083	0.042	0.125	0.167	**0.458**
AudioNet_LayerALL						
	Angry	Somber	Upbeat	Peaceful	Empowering	Stirring
Angry	**0.866**	0.015	0.015	0.000	0.015	0.090
Somber	0.011	**0.557**	0.034	0.250	0.000	0.148
Upbeat	0.000	0.000	**0.895**	0.000	0.070	0.035
Peaceful	0.000	0.143	0.000	**0.818**	0.000	0.039
Empowering	0.134	0.037	0.085	0.000	**0.732**	0.012
Stirring	0.133	0.067	0.100	0.000	0.133	**0.567**

Except for the overall accuracy, in the six classes, five classes' accuracy scores have been lift.

As seen in baseline's result, $\mathbf{C}_{Somber,Peaceful}$ and $\mathbf{C}_{Peaceful,Somber}$ are both not small. Since Somber and Peaceful are both less arousal, the network tend to be confused with each other. If more attention is put on valence related features, this phenomenon could be eased to some extent. After adding channel-wise attention mechanism, though accuracy for Somber has been reduced by 0.026, that for Peaceful has been improved by 0.192. This illustrates channel-wise attention mechanism's ability to re-weight and concentrate more on target-related feature maps.

As for Stirring, the baseline's accuracy score for which is the lowest. It is easily misclassified into Angry or Empowering because they are all more arousal. After adding the attention mechanism, the accuracy score for it has been improved.

4.3 Performance on Regression Task

Since the baseline has been proposed for song level emotion detection, rather than dynamic one, and there is no such type of dataset, we processed dynamic annotations in public dataset to generate the corresponding song level one, detailed in Sect. 3.2. In experiment, we conduct two set of experiments for arousal regression and valence regression.

Table 3. RMSE for the baseline and the proposed

Network	Arousal	Valence
AudioNet	0.301	**0.251**
AudioNet_LayerALL	**0.284**	0.253

As seen in Table 3, for arousal regression, the proposed performs better with a obviously smaller RMSE, while for valence regression, the baseline performs slightly better.

5 Conclusion

In this paper, channel-wise attention mechanism is introduced and designed to make the network focus more on the emotion related feature maps. Experiment results have verify the utility of the proposed method on both classification and regression tasks. In future work, we will concentrate more on the attention scheme, such as designing more sophisticate and accurate channel descriptors or introducing spatial attention mechanism.

Ackowlegement. This work was supported in part by National Key R&D Program of China (2019YFC1711800), NSFC (61671156).

References

1. Aljanaki, A., Yang, Y.H., Soleymani, M.: Developing a benchmark for emotional analysis of music. PloS One **12**(3), e0173392 (2017)
2. Chen, L., Zhang, H., Xiao, J., Nie, L., Shao, J., Liu, W., Chua, T.: SCA-CNN: spatial and channel-wise attention in convolutional networks for image captioning. In: 2017 IEEE Conference on Computer Vision and Pattern Recognition, CVPR 2017, Honolulu, HI, USA, 21–26 July 2017, pp. 6298–6306. IEEE Computer Society (2017). https://doi.org/10.1109/CVPR.2017.667

3. Choi, K., Fazekas, G., Sandler, M.B., Cho, K.: Transfer learning for music classification and regression tasks. In: Cunningham, S.J., Duan, Z., Hu, X., Turnbull, D. (eds.) Proceedings of the 18th International Society for Music Information Retrieval Conference, ISMIR 2017, Suzhou, China, 23–27 October 2017, pp. 141–149 (2017). https://ismir2017.smcnus.org/wp-content/uploads/2017/10/12_Paper.pdf
4. Corbetta, M., Shulman, G.L.: Control of goal-directed and stimulus-driven attention in the brain. Nat. Rev. Neurosci. **3**(3), 201–215 (2002)
5. Delbouys, R., Hennequin, R., Piccoli, F., Royo-Letelier, J., Moussallam, M.: Music mood detection based on audio and lyrics with deep neural net. In: Gómez, E., Hu, X., Humphrey, E., Benetos, E. (eds.) Proceedings of the 19th International Society for Music Information Retrieval Conference, ISMIR 2018, Paris, France, 23–27 September 2018, pp. 370–375 (2018). http://ismir2018.ircam.fr/doc/pdfs/99_Paper.pdf
6. Flexer, A., Schnitzer, D., Gasser, M., Widmer, G.: Playlist generation using start and end songs. In: Bello, J.P., Chew, E., Turnbull, D. (eds.) ISMIR 2008, 9th International Conference on Music Information Retrieval, Drexel University, Philadelphia, PA, USA, 14–18 September 2008, pp. 173–178 (2008). http://ismir2008.ismir.net/papers/ISMIR2008_143.pdf
7. Hu, J., Shen, L., Sun, G.: Squeeze-and-excitation networks. CoRR abs/1709.01507 (2017). http://arxiv.org/abs/1709.01507
8. Hu, X., Downie, J.S., Laurier, C., Bay, M., Ehmann, A.F.: The 2007 MIREX audio mood classification task: lessons learned. In: Bello, J.P., Chew, E., Turnbull, D. (eds.) ISMIR 2008, 9th International Conference on Music Information Retrieval, Drexel University, Philadelphia, PA, USA, 14–18 September 2008, pp. 462–467 (2008). http://ismir2008.ismir.net/papers/ISMIR2008_263.pdf
9. Kim, Y.E., Schmidt, E.M., Migneco, R., Morton, B.G., Richardson, P., Scott, J.J., Speck, J.A., Turnbull, D.: State of the art report: music emotion recognition: a state of the art review. In: Downie, J.S., Veltkamp, R.C. (eds.) Proceedings of the 11th International Society for Music Information Retrieval Conference, ISMIR 2010, Utrecht, Netherlands, 9–13 August 2010, pp. 255–266. International Society for Music Information Retrieval (2010). http://ismir2010.ismir.net/proceedings/ismir2010-45.pdf
10. Li, T., Ogihara, M.: Content-based music similarity search and emotion detection. In: 2004 IEEE International Conference on Acoustics, Speech, and Signal Processing, ICASSP 2004, Montreal, Quebec, Canada, 17–21 May 2004, pp. 705–708. IEEE (2004). https://doi.org/10.1109/ICASSP.2004.1327208
11. Liu, H., Fang, Y., Huang, Q.: Music emotion recognition using a variant of recurrent neural network. In: 2018 International Conference on Mathematics, Modeling, Simulation and Statistics Application (MMSSA 2018). Atlantis Press (2019)
12. Liu, X., Chen, Q., Wu, X., Liu, Y., Liu, Y.: CNN based music emotion classification. CoRR abs/1704.05665 (2017). http://arxiv.org/abs/1704.05665
13. Malik, M., Adavanne, S., Drossos, K., Virtanen, T., Ticha, D., Jarina, R.: Stacked convolutional and recurrent neural networks for music emotion recognition. CoRR abs/1706.02292 (2017). http://arxiv.org/abs/1706.02292
14. Mion, L., Poli, G.D.: Score-independent audio features for description of music expression. IEEE Trans. Speech Audio Process. **16**(2), 458–466 (2008). https://doi.org/10.1109/TASL.2007.913743

15. Panda, R., Malheiro, R., Paiva, R.P.: Musical texture and expressivity features for music emotion recognition. In: Gómez, E., Hu, X., Humphrey, E., Benetos, E. (eds.) Proceedings of the 19th International Society for Music Information Retrieval Conference, ISMIR 2018, Paris, France, 23–27 September 2018, pp. 383–391 (2018). http://ismir2018.ircam.fr/doc/pdfs/250_Paper.pdf
16. Robnik-Sikonja, M., Kononenko, I.: Theoretical and empirical analysis of ReliefF and RReliefF. Mach Learn **53**(1–2), 23–69 (2003). https://doi.org/10.1023/A:1025667309714
17. Russell, J.A.: A circumplex model of affect. J. Pers. Soc. Psychol. **39**(6), 1161 (1980)
18. Yang, X., Dong, Y., Li, J.: Review of data features-based music emotion recognition methods. Multimedia Syst. **24**(4), 365–389 (2018)
19. Yoon, S., Byun, S., Jung, K.: Multimodal speech emotion recognition using audio and text. In: 2018 IEEE Spoken Language Technology Workshop, SLT 2018, Athens, Greece, 18–21 December 2018, pp. 112–118. IEEE (2018). https://doi.org/10.1109/SLT.2018.8639583
20. You, Q., Jin, H., Wang, Z., Fang, C., Luo, J.: Image captioning with semantic attention. In: 2016 IEEE Conference on Computer Vision and Pattern Recognition, CVPR 2016, Las Vegas, NV, USA, 27–30 June 2016, pp. 4651–4659. IEEE Computer Society (2016). https://doi.org/10.1109/CVPR.2016.503
21. Zeiler, M.D., Fergus, R.: Visualizing and understanding convolutional networks. In: Fleet, D.J., Pajdla, T., Schiele, B., Tuytelaars, T. (eds.) Computer Vision - ECCV 2014 - 13th European Conference, Zurich, Switzerland, 6–12 September 2014, Proceedings, Part I, Lecture Notes in Computer Science, vol. 8689, pp. 818–833. Springer (2014). https://doi.org/10.1007/978-3-319-10590-1_53
22. Zhang, S., Tian, Q., Jiang, S., Huang, Q., Gao, W.: Affective MTV analysis based on arousal and valence features. In: Proceedings of the 2008 IEEE International Conference on Multimedia and Expo, ICME 2008, 23–26 June 2008, Hannover, Germany, pp. 1369–1372. IEEE Computer Society (2008). https://doi.org/10.1109/ICME.2008.4607698
23. Zhang, Y., Li, K., Li, K., Wang, L., Zhong, B., Fu, Y.: Image super-resolution using very deep residual channel attention networks. In: Ferrari, V., Hebert, M., Sminchisescu, C., Weiss, Y. (eds.) Computer Vision - ECCV 2018 - 15th European Conference, Munich, Germany, 8–14 September 2018, Proceedings, Part VII, Lecture Notes in Computer Science, vol. 11211, pp. 294–310. Springer (2018). https://doi.org/10.1007/978-3-030-01234-2_18

Symbolic Melody Phrase Segmentation Using Neural Network with Conditional Random Field

Yixiao Zhang and Gus Xia(✉)

Music X Lab, NYU Shanghai, Shanghai 200122, China
{yixiao.zhang,gxia}@nyu.edu

Abstract. Automatic melodic phrase detection and segmentation is a classical task of content-based music information retrieval and also a key problem of automatic music structure analysis. In this paper, we apply neural network architectures with conditional random field (CRF) to produce satisfactory melodic phrase segmentation. To tackle the problem of the *sparse labelling* problem of data, we design two tailored label-engineering techniques with corresponding training techniques for different neural networks. We compare the performance of the traditional model on the public data set Essen Folksong Database. The experimental results show that the performance of the model using the neural CRF architecture far exceeds that of the traditional method. The results of ablation experiments on Essen Folksong Database and POP909 dataset show that the improvement of performances mainly comes from the introduction of CRF structure. Besides, our labelling techniques also improve model performance and make training process more robust (Codes and data are available at https://github.com/ldzhangyx/music-melody-segmentation-using-neural-CRF).

Keywords: Music segmentation · Conditional random field · Music information retrieval

1 Introduction

Automated melodic phrase detection and segmentation is a classical task in content-based music information retrieval (MIR). It is also the key step towards automated music structure analysis, which is useful for many computer-music applications, such as structured automated composition [21], music databases [11], and query-by-humming [10]. However, current solutions for melodic phrase detection cannot yet satisfy practical requirements, especially for symbolic music representation. To be specific, rule-based methods, in general, rely on theme repetitions, long notes and rests, and hence are unstable when dealing with music with large variations; traditional machine-learning methods rely on manually-designed features and very difficult to capture useful music context information for boundary detection.

X. Shao et al. (Eds.): CSMT 2020, LNEE 761, pp. 55–65, 2021.
https://doi.org/10.1007/978-981-16-1649-5_5

On the other hand, many neural network architectures have recently achieved quite promising results in various domains, including representation learning [2], computer vision [14], natural language processing [6], autonomous driving [3]. Analogous to the problem of adding punctuation to a list of characters to form sentences, in this paper, we are primarily interested in labelling the begin and end of a phrase given a sequence of notes. Since supervised neural networks can use the back-propagation mechanism to automatically identify the crucial music-context related features in various ways, we experiment a combination of existing neural networks and probabilistic graphical models to solve the task, including convolutional neural network with the conditional random field (CNN-CRF) and bi-directional long short-term memory with the conditional random field (Bi-LSTM-CRF).

The main issue of applying deep-learning methods to phrase detection is the *sparse labelling* of the training sets. To address this issue, we:

- Contributed two label engineering techniques to solve the *sparse labelling* problem that hinders the use of sequential decision-making neural networks;
- Combined the labelling techniques with our deep learning models, which considers both implicit and explicit relationships between labels to detect phrase boundaries in symbolic representations of music;
- Conducted a quantitative evaluation of the performance of the proposed models for the task.

All models are trained and tested on Essen Folksong Dataset (EFSC) and POP909 dataset. Experiment results show that the Bi-LSTM-CRF architecture performs the best, being able to offer finer segmentation and faster to train, while CNNs, Bi-LSTM-CNNs and CNN-CRFs are acceptable alternatives.

In the following sections, we discuss the related work in Sect. 2 and present the methodology in Sect. 3 followed by the experimental results and analysis in Sect. 4. Finally, we conclude our paper with some reflections and possible directions for future works in Sect. 5.

2 Related Work

Melody phrase segmentation task has attracted researchers for decades. Foote [8] first proposed a method to visualizes the structure of music audios by self-similarity matrices. Later, by measuring changes in local self-similarity, Foote [9] developed a rule-based boundary detection method for automatic audio segmentation. Kaiser and Sikora [13] further proposed a method that applies non-negative matrix factorization to self-similarity matrices to produces two factorization products, based on which the structure boundaries can be derived. Other methods include Hidden Markov Model [1], decision tree [27] and clustering [17] for audio representation and Local Boundary Detection Model (LBDM) [4], Grouper [26], IDyOM [22] and Restricted Boltzmann Machines [15] for symbolic representation. Most rule-based methods follow one or more rules in Generative Theory of Tonal Music (GTTM) [16] system. For example, the LBDM model follows similar rules as GPR 3, while the Grouper model uses GPR 2 and PSPR 1 rules. Compared with deep learning model, these rule-based methods severely rely on given rules, lacking the ability of not only discovering new relationships of data but also updating and optimizing rules for specific data.

Recently, neural networks are applied to tackle the task. Ullrich et al. [28] used Convolutional Neural Networks to analyze music structure, which is more relevant to our work. The model takes a spectrogram as an input and outputs the probability of phrase boundary for each spectrum, following by which peak-picking and thresholding algorithms are applied to post-process the result. Our study also considers the problem of phrase segmentation but focus on symbolic representations. This model tries to use neural network to solve the problem, but the simple model structure limits its performance. The method of using CNN only is treated the baseline method. We implemented a CNN segmentation model in the ablation study to represent the performance of this model.

Different from traditional methods which heavily rely on rests and long notes [25], our system is learning-based and takes into account more music contexts, which performs better on tackling the problem of "jump-phrases" of those phrases go across the temporal, which typically a severe challenge for all methods which only detect boundaries ignoring the content of melodies. Moreover, rather than using rule-based post-processing methods, we used a CRF and combined it with deep learning architectures, making the system end-to-end. By applying neural network and CRF, our model on segmentation tasks performs significantly better than previous methods.

Neural conditional random field has been widely used in name entity recognition tasks [7,12,20] and image segmentation tasks [19,23]. For the first time, we tried to apply the neural CRF structure to the phrase-level segmentation task of the entire music length data, which requires more domain-specific knowledge as inductive biases.

3 Method

We introduce the problem definition in Sect. 3.1 and present the neural network with CRF models in Sect. 3.2. Then, we discuss two label engineering techniques for model training in Sect. 3.3. Finally, we introduce loss functions used for training can be found in Sect. 3.4.

3.1 Problem Definition

In this section, we formally define our problem and introduce our data representation in detail. We denote $\mathbf{X} \in \mathbb{R}^T$ as the random variable over music sequence to be labeled and $\mathbf{Y} \in \mathbb{N}^T$ as the random variable over the space of all valid label sequences, where T is the length of the sequence. A specific music phrase is denoted as $\{x_i\} = \mathbf{x} \sim F_X$, and $\hat{y_i} = \phi(x_i)$ is the predicted label generated by the model ϕ, where $i \in [1, T]$. We use y_i^* as the corresponding ground truth label.

Our goal is to construct the conditional probability $P(Y|X)$, which is approximated by $p(y|x) = \phi_\theta(x)$, (x, y samples from the dataset $\mathcal{D}$). The model $\phi_\theta(x)$ is optimized using maximum likelihood estimation by finding $\underset{\theta \in \Omega}{\operatorname{argmax}} \mathcal{L}(y*, \phi_\theta(x))$. While in practice we approach the problem by performing empirical risk minimization $\underset{\theta \in \Omega}{\operatorname{argmin}} \frac{1}{N} \sum_{i=1}^{N} Loss(y_i*, \phi_\theta(x_i))$, where Ω is the parameter space.

3.2 Neural Networks with Conditional Random Field

We use CNN or Bidirectional LSTM to transform and encode input melody. We apply skip-connection between adjunct layers to avoid the problem of gradient exploration. Formally, in the layer n, hidden states are calculated as below. For the LSTM encoder:

$$\begin{aligned} h_n &= \text{Bi-LSTM}(f(h_{n-1}) + f(h_{n-2})) \\ &= [\overrightarrow{\text{LSTM}}(f(h_{n-1}) + f(h_{n-2})), \overleftarrow{\text{LSTM}}(f(h_{n-1}) + f(h_{n-2}))] \end{aligned} \tag{1}$$

and for the CNN encoder:

$$h_n = \text{CNN}(f(h_{n-1}) + f(h_{n-2})) \tag{2}$$

where function $f(\cdot)$ is non-linear activate function. Particularly, $f(\cdot) = \text{ReLU}(\cdot)$. The entire network takes X as input, therefore $h_0 = X$. Skip connection mechanism is not applied in the first two layers. Then, hidden states are transformed back to T-dims vector by a linear transformation $\widetilde{x} = Wh + b$.

We consider the matrix of scores $f_\theta(\widetilde{x})$ are the output of the network. For the i-th tag, at the t-th word, the element $f_{\theta i,t}$ is the score output by the network with parameter θ. A transition score $A_{i,j}$ models the transition from i-th to j-th for a pair of consecutive time steps. We denote the new parameters for the network as $\widetilde{\theta} = \theta \bigcup \{A_{i,j} \forall i, j\}$. The score of a sentence $\widetilde{x}$ along with a path of tags i is given by:

$$s(\widetilde{x}, i, \widetilde{\theta}) = \sum_{t=1}^{T} (A_{i_{t-1}, i_t} + f_{\theta i_t, t}) \tag{3}$$

The dynamic programming algorithm are able to compute $A_{i,j}$ and optimal tag sequences for inference. Finally, the entire network can be end-to-end trained.

3.3 Two Label Engineering Techniques

Most previous works posed the musical phrase segmentation task naturally as a binary classification problem and thus deployed the binary labelling scheme. Consequently, dataset has highly-imbalanced label distribution, which makes the training process much harder for many neural network architectures. To solve the *sparse labelling* problem, we propose two alternative label engineering techniques, while keeping the 0–1 labeled dataset for the training of our baseline methods:

- One is named as *exponential-decay label*, where we assign value 1 to the start of a music phrase, starting from where the value decays exponentially to $\frac{1}{2}$, $\frac{1}{4}$ till it reaches to the middle of a sentence before it goes up with the same rate till the start of next phrase;
- the other is named as *linear-ascend label*, where we assign to a note the value of that note's numbered position of the phrase it is in. Notes not in any phrases will be assigned label 0.

A visualization of these three types of labelling is provided in Fig. 1. This two labelling methods clearly avoid the label imbalance problem, and in particular, the linear ascending label can be used to train neural network with CRF as its final layer since the state transitional matrix of CRF requires a discrete label space.

3.4 Loss Functions

We introduce three training objectives correlated to different labelling techniques.

Loss Function for Binary Labels. We adopt the cross entropy loss for binary classification to form this loss function. To deal the highly imbalanced distribution of label, we introduce into the loss function a large weight factor α for label 1:

$$loss(y^\star, \hat{y}) = \sum_i [-\alpha \ y_i^\star \log(\hat{y_{i,1}}) - (1 - y_i^\star) \log(1 - \hat{y_{i,0}})] \quad (4)$$

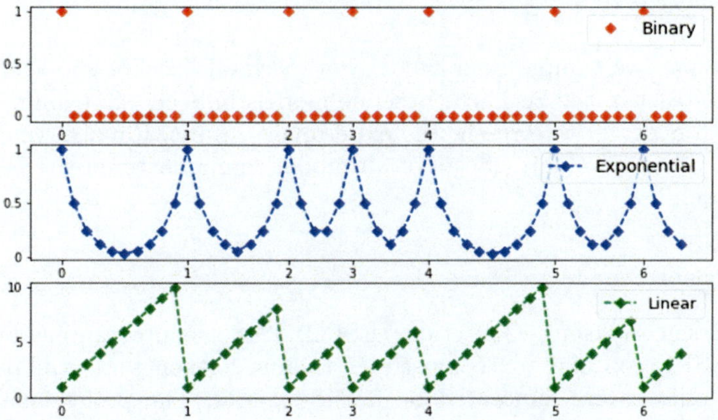

Fig. 1. The original binary labels are displayed in the first row, and the "exponential-decay label" and "linear-ascend label" are shown in the second and third row.

Loss Function for Exponential-Decay Labels. Inspired by focal loss [18], we introduce a re-scaling factor into Mean Square Error (MSE) loss to help our model better fit the exponential-decay labels:

$$loss(y^\star, \hat{y}) = \frac{1}{n} \sum z_i \tag{5}$$

$$z_i = \begin{cases} \alpha \left(\hat{y}_i - y_i^\star\right)^2 & \text{if } y_i^\star = 1 \; and \; |\hat{y}_i - y_i^\star| < 1 \\ \frac{1}{2} \left(\hat{y}_i - y_i^\star\right)^2 & \text{if } y_i^\star \neq 1 \; and \; |\hat{y}_i - y_i^\star| < 1 \\ |\hat{y}_i - y_i^\star| - \frac{1}{2} & \text{otherwise} \end{cases} \tag{6}$$

where α is the penalty rate.

Loss Function for Linear-Ascend Labels. Since only the training of CRF variants involves the use of linear-ascend label, we present here the standard loss function of CRF that seeks to maximize the negative log-likelihood of the ground-truth sequence label:

$$\begin{aligned} loss(x, y^\star) = & \sum_{\hat{y}} \sum_{i=0}^{n} \log(\mathcal{L}[x_i|\hat{y}_i)T(\hat{y}_i|\hat{y_{i-1}})] \\ & - \sum_{i=1}^{n} \log[\mathcal{L}(x_i|y_i^\star)T(y_i^\star|y_{i-1}^\star)] \end{aligned} \tag{7}$$

Note that during training, CRF does not need to produce any prediction sequence $\hat{y}$. Instead, it works with all possible label sequences given x.

4 Experiments

In this section, we compare our model woth several baseline models. We introduce the dataset in the Sect. 4.1, baseline models in Sect. 4.2, training details in Sect. 4.3. In Sect. 4.4 we present the experiment results. Finally, we conduct an ablation study to evaluate effects of all model components on the performance of our model.

4.1 Dataset

We use Essen Folksong Collection (ESFC) [24] as our training and testing dataset. ESFC contains 6,236 mostly Germanic folksongs in symbolic format, where all phrases are annotated by music experts. The professional labelling of the data alleviates the problem of ill-definition of the melody segmentation task to a certain extent, making it usable in engineering. Hence, ESFC has been widely used in testing segmentation models and provides a common basis for different models to be compared. We randomly split the dataset (at song-level) into training set (90%) and test set (10%).

We employ data augmentation during the training stage. In particular, We make the length of silence and the length of notes randomly offset by 0.3 and 0.5 respectively, and transpose music to all 12 keys. Since our model attempts to enhance the performance of segmentation tasks by understanding the context of the melody, it is necessary to perform data augmentation on the tonality of music.

Besides, we also use POP909 dataset [29] for further experiments, which contains about 1000 well-known pop songs with melody and phrase labels. We apply the same augmentation technique to the dataset.

4.2 Baselines

We choose **Pause** model, **LBDM** [4], **Grouper** [26] and the previous state-of-the-art model **ΔIOI** [5] as our baselines. The **Pause** model is a simple model which only considers silence tokens as phrase boundaries. The pause model cannot tackle the problem of jump-phrases. **LBDM** is a rule-based method, which consists of a change rule and a proximity rule operating over melodic information that encode pitch, intervals and rests. **Grouper** uses 3 PSPR rules to asses the existence of segment boundaries, which are defined in GTTM system. In **ΔIOI**, boundaries are selected by calculating differences between successive intervals. It has a compound version that trains a meta classifier to improve performance given rule-based results and ΔIOI results.

4.3 Training Settings

For both CNN-CRF and LSTM-CRF models, we set batch size to 32 and pad sequence length to 120. When the linear labelling is used, we set a total number $k = 32$ for different label tags, which is larger than the maximum length of one single phrase, while $k = 4$ When we use binary labels. the n We set the learning rate to 0.01 with a scheduler that scales down the learning rate by 0.75 at the end of epoch. When initializing the network, we put a large negative number at $A_{i,j}$, where A is the transition matrix of the final CRF layer, making it illegal for a sequence to jump from label j to label i and enforcing the network only considers valid prediction.

For the Bi-LSTM-CRF model, 7 Bi-LSTM layers are stacked along with skip connection. We set hidden size to 512; For the CNN-CRF model, 5-layer CNN are applied and we set kernels to 3, 3, 3 and 5 respectively. Hidden size of each layer is 512.

4.4 Results and Analysis

We evaluate our models by comparing different models in terms of segmentation accuracy using F1 score. Table 1 shows the results where we see that both Bi-LSTM-CRF model and CNN-CRF perform much better than previous methods.

We also perform an ablation study to analyze the contribution of each part of the model to performance. We remove the CRF module from the model,

Table 1. Evaluation results on melody segmentation task.

Model	F-1 score
ΔIOI	0.58
Pause	0.60
LBDM	0.65
Grouper without meter	0.66
Grouper with meter	0.74
ΔIOI (Compound)	0.75
CNN-CRF (Linear)	0.82
Bi-LSTM-CRF (Linear)	**0.84**

leaving a separate Bi-LSTM or CNN network for training. We additionally set up a stacking network of LSTM-CNN for comparison. In the experiments we use different types of labels. For the neural CRF model, we use linear labels; for other models, we use exponential labels. Table 2 shows the result of ablation study. Ablation study are performed on EFSC and POP909 dataset.

Table 2. Evaluation results of the ablation study. Both CRF architecture and label engineering improve the model performance.

Model	Binary	Linear	Exponential
Bi-LSTM	0.75	–	0.73
CNN	0.74	–	**0.76**
Bi-LSTM-CNN	0.76	–	0.74
CNN-CRF	0.81	**0.82**	–
Bi-LSTM-CRF	0.82	**0.84**	–

We observed that the introduction of the CRF structure is accompanied by a great performance improvement; even if the CRF structure is not used, the deep neural network alone exceeds the best performance of the existing model.

Experiment results also shows that the linear label engineering techniques slightly improve the performances and both linear labels and exponential labels make the training process of the CRF models more robust.

Another observation is that time series models, once properly set up, can perform better than the CNN variants for this particular task. This can be explained by time series models' ability to capture long-term dependencies in theory, yet the performance boost comes at the expense of increased training time. Time series models typically take a longer time to train under the settings described in Sect. 4.4. Bi-LSTM-CNN and Bi-LSTM-CRF models take the longest time to train. While CNN-CRF can be trained much faster than Bi-LSTM-CRF and

Bi-LSTM-CNN, the training of it is still slower than the training of the CNN model. In general, the CNN-CRF model achieves the best balance between time cost and performance.

A further visualization of each model's prediction agrees with the performance ranking in Table 2. For brevity, we include in Fig. 2 only the corresponding ground-truth label of phrase boundaries and predictions given by the CNN, Bi-LSTM-CNN and CNN-CRF models on a music sample from the held-out test set, with red lines indicating there are phrase boundaries at these position according to the ground-truth label, and blue, yellow and orange lines indicating phrase boundary predictions by CNN-CRF/Bi-LSTM-CRF, Bi-LSTM-CNN and CNN models respectively, where Bi-LSTM-CRF gives the same result as CNN-CRF. As shown in Fig. 2, CNN model is only able to find half of the phrase boundaries, while Bi-LSTM-CNN/CNN-CRF model manages to find most of the phrase boundaries, missing only one place. CNN-CRF model not only successfully identifies all phrase boundaries according to the ground-truth label, but also marks one more place as possible phrase boundary point.

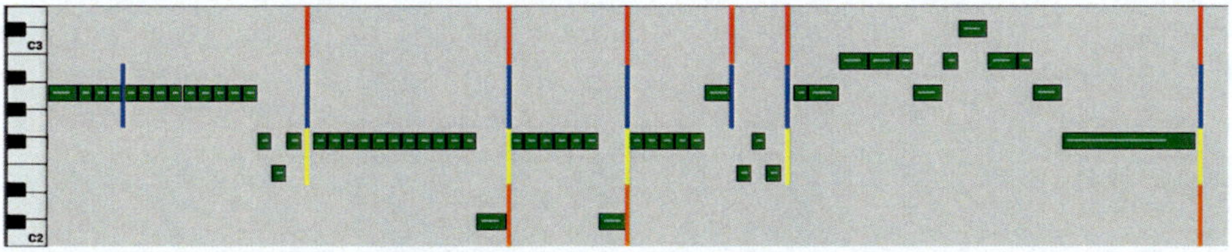

Fig. 2. An example of segmentation results by different models. Red: Ground Truth; Blue: CNN-CRF/Bi-LSTM-CRF (same results); Yellow: Bi-LSTM-CNN; Orange: CNN.

In addition, our best-performing models are able to develop its own understanding of music, which make themselves give reasonable segmentation results. Currently, the melodic segmentation task is actually an ill-defined problem as humans have rather high agreement on all others, and improving the ability of music understanding of models is beneficial to alleviate the problem. Figure 3 visualizes the phrase boundaries the Bi-LSTM-CRF/CNN-CRF model predicts on another music sample from our test set and the corresponding ground-truth label, following the same notation used in Fig. 2. There are places that can be deemed as the start or the end of a phrase, but the ground truth label chooses not to mark these places as boundaries. Our neural CRF models can not only successfully predict where the ground truth labels think a phrase begins or ends, but identify these places as boundaries and produce a finer phrase segmentation.

5 Conclusion

We introduce in this paper a set of deep learning architectures and two label engineering techniques for the symbolic music phrase segmentation task. Experiment results indicates the effectiveness of our label engineering techniques. While

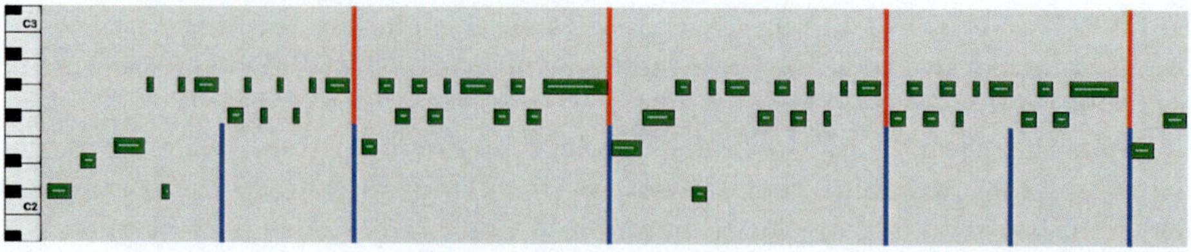

Fig. 3. Prediction results. Red: Ground Truth; Blue: Bi-LSTM-CRF/CNN-CRF. It reveals that understanding music content is helpful to make reasonable segmentation although those are not in ground truth.

all models can yield satisfactory phrase segmentation, combining CRF with deep neural networks dramatically improves the performance of our models, as CRF explicitly characterizes the relation among labels. In the feature, we plan to explore the segmentation task with BERT-CRF architecture, and apply neural CRF models to more sophistic tasks. e.g. music structure analysis.

Acknowledgement. The preliminary partial results of this work are available at https://arxiv.org/abs/1811.05688.

References

1. Aucouturier, J.J., Sandler, M.: Segmentation of musical signals using hidden Markov models. In: Preprints-Audio Engineering Society (2001)
2. Bengio, Y., Courville, A., Vincent, P.: Unsupervised feature learning and deep learning: a review and new perspectives. CoRR abs/1206.5538 (2012). http://arxiv.org/abs/1206.5538
3. Bojarski, M., Del Testa, D., Dworakowski, D., Firner, B., Flepp, B., Goyal, P., Jackel, L., Monfort, M., Muller, U., Zhang, J., et al.: End to end learning for self-driving cars. arXiv preprint arXiv:160407316 (2016)
4. Cambouropoulos, E.: The local boundary detection model (LBDM) and its application in the study of expressive timing. In: ICMC (2001)
5. Cenkerová, Z., Hartmann, M., Toiviainen, P.: Crossing phrase boundaries in music. In: Proceedings of the Sound and Music Computing Conferences (2018)
6. Cho, K., van Merrienboer, B., Gülçehre, Ç., Bougares, F., Schwenk, H., Bengio, Y.: Learning phrase representations using RNN encoder-decoder for statistical machine translation. CoRR abs/1406.1078 (2014). http://arxiv.org/abs/1406.1078
7. Dong, C., Zhang, J., Zong, C., Hattori, M., Di, H.: Character-based LSTM-CRF with radical-level features for Chinese named entity recognition. In: Natural Language Understanding and Intelligent Applications, pp. 239–250. Springer (2016)
8. Foote, J.: Visualizing music and audio using self-similarity. In: Proceedings of the Seventh ACM International Conference on Multimedia (Part 1), pp. 77–80. ACM (1999)
9. Foote, J.: Automatic audio segmentation using a measure of audio novelty. In: 2000 IEEE International Conference on Multimedia and Expo. ICME 2000, vol. 1, pp. 452–455. IEEE (2000)

10. Ghias, A., Logan, J., Chamberlin, D., Smith, B.: Query by humming: musical information retrieval in an audio database. In: Proceedings of the Third ACM International Conference on Multimedia, pp. 231–236. ACM (1995)
11. Hashida, M., Matsui, T., Katayose, H.: A new music database describing deviation information of performance expressions. In: ISMIR, pp. 489–494 (2008)
12. Huang, Z., Xu, W., Yu, K.: Bidirectional LSTM-CRF models for sequence tagging. arXiv preprint arXiv:150801991 (2015)
13. Kaiser, F., Sikora, T.: Music structure discovery in popular music using non-negative matrix factorization. In: ISMIR, pp. 429–434 (2010)
14. Krizhevsky, A., Sutskever, I., Hinton, G.: ImageNet classification with deep convolutional neural networks. In: Advances in Neural Information Processing Systems, pp. 1097–1105 (2012)
15. Lattner, S., Grachten, M., Agres, K., Chacón, C.: Probabilistic segmentation of musical sequences using restricted Boltzmann machines. In: Mathematics and Computation in Music, pp. 323–334. Springer (2015)
16. Lerdahl, F., Jackendoff, R., et al.: A Generative Theory of Tonal Music, vol. 1996. MIT Press, Cambridge (1983)
17. Levy, M., Sandler, M.: Structural segmentation of musical audio by constrained clustering. IEEE Trans. Audio Speech Lang. Process. **16**(2), 318–326 (2008)
18. Lin, T., Goyal, P., Girshick, R., He, K., Dollár, P.: Focal loss for dense object detection. CoRR abs/1708.02002 (2017). http://arxiv.org/abs/1708.02002
19. Liu, F., Lin, G., Shen, C.: CRF learning with CNN features for image segmentation. Pattern Recogn. **48**(10), 2983–2992 (2015)
20. Ma, X., Hovy, E.: End-to-end sequence labeling via bi-directional LSTM-CNNS-CRF. arXiv preprint arXiv:160301354 (2016)
21. Nierhaus, G.: Algorithmic Composition: Paradigms of Automated Music Generation. Springer, Heidelberg (2009)
22. Pearce, M., Müllensiefen, D., Wiggins, G.: Melodic grouping in music information retrieval: new methods and applications. In: Advances in Music Information Retrieval, pp 364–388. Springer (2010)
23. Roy, A., Todorovic, S.: Combining bottom-up, top-down, and smoothness cues for weakly supervised image segmentation. In: Proceedings of the IEEE Conference on Computer Vision and Pattern Recognition, pp. 3529–3538 (2017)
24. Schaffrath, H.: The Essen folksong collection. Database containing 6 (1995)
25. Takasu, A., Yanase, T., Kanazawa, T., Adachi, J.: Music structure analysis and its application to theme phrase extraction. In: International Conference on Theory and Practice of Digital Libraries, pp. 92–105. Springer (1999)
26. Temperley, D.: The Cognition of Basic Musical Structures. MIT Press, Cambridge (2004)
27. Turnbull, D., Lanckriet, G., Pampalk, E., Goto, M.: A supervised approach for detecting boundaries in music using difference features and boosting. In: ISMIR, pp. 51–54 (2007)
28. Ullrich, K., Schlüter, J., Grill, T.: Boundary detection in music structure analysis using convolutional neural networks. In: ISMIR, pp. 417–422 (2014)
29. Wang, Z., Chen, K., Jiang, J., Zhang, Y., Xu, M., Dai, S., Bin, G., Xia, G.: POP909: a pop-song dataset for music arrangement generation. In: Proceedings of 21st International Conference on Music Information Retrieval, ISMIR (2020)

Automatic Recognition of Basic Guzheng Fingering Techniques

Hailei Ding, Hao Zhang, Bingqiang Yan, Junjun Jiang, Min Huang(✉), and Zhongzhe Xiao(✉)

School of Optoelectronic Science and Engineering, Soochow University, Suzhou 215006, Jiangsu, China
{hmin,xiaozhongzhe}@suda.edu.cn

Abstract. Automatic recognition of six selected basic Guzheng fingering techniques is performed in this paper. The audio samples considered in this work are cut into segments of single notes to emphasize the property of each fingering. Due to limited scale of audio samples, traditional machine learning methods are used in the automatic recognition instead of deep learning ones. The RMS energy and MFCCs are proved to be the most effective parameters in presenting Guzheng fingerings. The accuracy in the recognition of the six selected fingerings reaches up to 90.73% with Random Forest, where the "Yao" achieves perfect recognition of 100%.

Keywords: Guzheng · Fingering recognition

1 Introduction

We aim to perform an automatic recognition of basic Guzheng fingering techniques in this work. Guzheng, one of important Chinese traditional musical instruments, has a history of more than 2500 years, which can be traced back to as early as the Warring States Period [1]. It is also known as Qinzheng, Hanzheng, Yaozheng or Luanzheng in the history. Its shape, number of strings, and the mode of tone settings are all developing overtime. There were 5 strings at the earliest, 13 strings in the Tang and Song Dynasties, and later increased to 16, 18, 21, and 25 strings. The common modern model Guzheng is specified as S21-163 as shown in Fig. 1. The body of the Guzheng is a rectangular wooden sound box, and it is composed of front panel, strings, string nails, Yueshan, sound outlet, bottom plate, etc. "S" in the model name stands for S-shaped Back Yueshan, 163 represents the length of the Guzheng is about 163 cm, and 21 represents the number of Guzheng strings. Guzheng may also appear with straight back Yueshan and sometimes with shorter lengths for portable purpose (mini Guzheng or half Zheng).

With 21 strings, Guzheng covers 4 octaves. Although it has only 5 tones (do, re, mi, sol, la) in each octave, the strings are normally tuned with frequencies according to twelve-tone equal temperament, to ensure its precision over the wide

X. Shao et al. (Eds.): CSMT 2020, LNEE 761, pp. 66–77, 2021.
https://doi.org/10.1007/978-981-16-1649-5_6

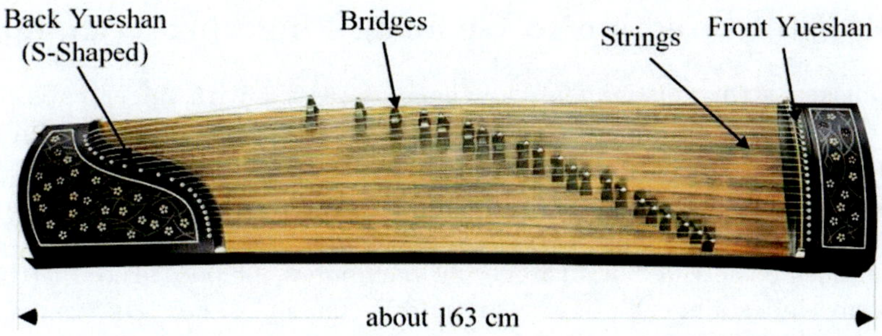

Fig. 1. Illustration of a Guzheng

range. The expressiveness of Guzheng is presented by various playing techniques, including both right hand fingering and left hand fingering techniques. In this paper, we focus on an automatic approach of Guzheng fingering recognition, aiming better analysis of Guzheng performing, and auxiliary teaching of Guzheng playing.

Until currently, research work of Guzheng fingering still focused in the playing executions [2–4], while very few study was implemented with the rapidly developing artificial intelligent approaches. The most relative work on automatic detection of Guzheng was a music-to-score alignment system that can mark out the errors in Guzheng playing [5]. Other close work, on Chinese traditional plucked instruments but not Guzheng, normally focused on music synthesis, such as Pipa and Qin [6,7]. For the automatic fingering recognition task, the most similar work lies in the field of instruments recognition, where both tasks concerning the timbre of the music. In instruments recognition, a number of acoustic features are investigated, including temporal features, spectral features, cepstral features, *etc.* [8]. A various of algorithms, including signal processing, feature processing, and machine learning classifiers, such as empirical mode decomposition (EMD), independent component analysis (ICA), HMM, K-NN, *etc.* [9–13]. We assume that the fingering recognition is also a timbre recognition task that similar to instruments recognition.

In this work, samples of single notes of Guzheng from 6 selected fingerings are collected, and with proper acoustic features and machine learning methods, we perform an automatic recognition of basic Guzheng fingerings. The aim of this work is to make accurate recognition of Guzheng fingering, and further automatically evaluate the playing quality of certain fingerings in our future work, to make it possible to build a tool for better comprehending Guzheng music and Guzheng playing. A possible application is to make an auxiliary teaching system of Guzheng playing, to contribute to the inheritance and promotion of this elegant Chinese traditional instrument.

The rest of the paper is organized as follows. In Sect. 2, we briefly introduce several basic Guzheng fingering techniques; the automatic recognition and analysis to the results are performed in Sect. 3; Sect. 4 concludes the paper and presents the future work.

2 Brief Introduction to Guzheng Fingering Techniques

As a plucked string instrument, Guzheng is played with finger caps on both hands. The right hand fingerings are responsible for the main melody in playing with a various of techniques, and the left hand fingerings are mainly for the glissando or vibrato effects by pressing the strings to the left of the bridges, or for accompaniment purpose with similar techniques to the right hand.

The most basic right hand fingering techniques include "Tuo/Pi (thumb)", "Mo/Tiao (index finger)", "Gou/Ti (middle finger)", and "Da (ring finger)", which are all played by quickly hitting a single string only once with one finger toward or outward the direction of palm [3].

Playing with multiple strings or multiple times is a very important way to enrich Guzheng playing techniques. For example, combinations of two strings are very commonly used in Guzheng music, including "Dacuo", which is combined with "Tuo" and "Gou" over an octave, and "Xiaocuo", which is combined with "Tuo" and "Mo" over a range smaller than an octave. As an instrument with a large number of strings, there are also several fingering techniques that hit multiple strings at a time, such as "Huazhi" and "Guazou". A string can be also hit multiple times over a short time to get an effect as a continuous sound, such as "Yao (Shaking)", which is a traditional Guzheng fingering technique, and "Lunzhi", which is adopted from Pipa playing. "Yao" refers to hit a certain string with one finger quickly and repeatedly at a stable speed, normally with the thumb, that is to say, to play "Tuo" and "Pi" alternatively and quickly.

More expressive Guzheng fingering techniques need the help of the left hand. Glissando of Guzheng is realized by pressing the left side of the string to enhance the tension on the string to get a higher tone. The tone after pressing is usually with a distance to its original tone of major second or minor third, according to different strings. If the string is pressed before hitting the string and released later, the fingering is called "Xiahua"; if the string is first hit and pressed later, the fingering is called "Shanghua". If the string is pressed and released repeatedly to a certain degree (normally not so deep as in "Shanghua" or "Xiahua"), we can get the effect of vibrato, called "Chanyin". Another fingering technique with the help of left hand is "Fanyin (overtone)", which is played by touching the middle point of string with the skin of the left little finger simultaneously with the right hand finger hitting the string, to make a crystal sound.

Besides these fingering techniques, there are also a lot of other Guzheng fingerings, including some special fingerings depending on different genres, such as "Youzhi" of Henan genre.

3 Automatic Fingering Recognition and Analysis

In this section, we make an experimental investigation on the automatic recognition of six selected basic Guzheng fingering techniques. Limited to the scale of Guzheng audio samples in our work, traditional machine learning methods are used instead of deep learning ones to avoid problems such as overfitting. Several different methods are used for comparison.

3.1 Experimental Setting

We first clarify the types of fingering, the features and the classifiers used in this Guzheng fingering recognition work.

Fingering Selection. As there are many fingering techniques of Guzheng, we do not aim to cover all the fingering techniques in this primary work. We select 6 basic fingerings which are used most commonly for the automatic recognition. All 6 selected fingerings are played on single string. The techniques over multiple strings, such as "Huazhi" or "Guazou", will present a clear pattern with multiple hitting points of string and rapid pitch changing, are temporarily excluded from this work.

First, the most basic techniques that hit a single string only once should be considered in the study, including "Tuo/Pi", "Gou/Ti", "Mo/Tiao", and "Da". These fingerings have some similarities between each other, and are considered together in this work. We use the term "Gou" for all these most basic fingering in the following sections.

Then, three fingerings that modify the pitch to some degree with the pressing of left hand are selected, including "Shanghua", "Xiahua", and "Chanyin (Vibrato)". In these 3 fingerings, the index finger, middle finger, and ring finger of the left hand press the left side of the bridge of the string being played, as shown in Fig. 2(a). "Shanghua" and "Xiahua" present a moving of pitch with a scale of major second or minor third during a tone, and "Chanyin" presents a slight continuous changing of pitch around a base level.

Another fingering with special timbre, "Fanyin", is also selected. In playing of "Fanyin", we touch the harmonic point of the corresponding string, normally the half point of the effective vibration length, with the ring finger lightly and rapidly, as shown in Fig. 2(b). The "Fanyin" in Guzheng presents a clear crystal sound, where the second harmony energy is dominant in the spectrum. It can be used in Guzheng music to mimic the effects of percussion instruments such as bells and drums.

The sixth selected fingering is "Yao", which hits the string back and forth quickly and repeatedly to keep the sound duration and the continuity of the music [14]. The "Yao" is normally played with the thumb, and sometimes the little finger of the right hand can be used as a supporting point, as shown in Fig. 2(c). It is a very important technology in the Guzheng playing technique, and quite different with the other selected fingerings in this work.

We collected samples of the above mentioned six fingerings to make a dataset for the automatic fingering recognition experiments. There are two sources of samples. Some samples come from the Guzheng part of CCMusic developed by China Conservatory of Music in 2019 [15], where the samples were played by professional performers, and other samples come from fingering exercises played by beginners in learning. The samples from both professional performers and beginners ensure the diversity of playing. There are totally 368 samples collected on these 6 fingerings, all types with almost balanced number.

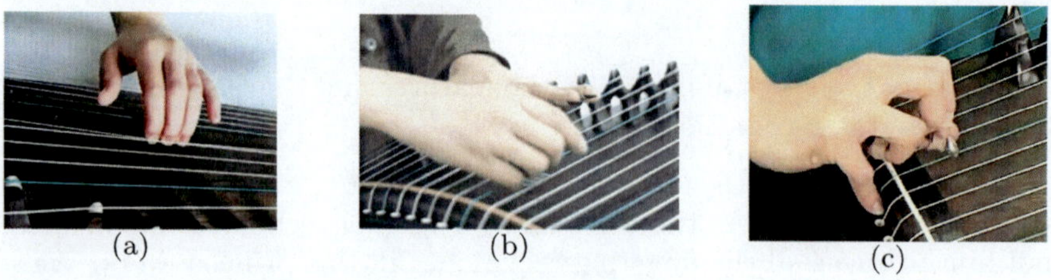

Fig. 2. Illustration of typical Guzheng fingerings (a) Pressing strings of "Chanyin", "Shanghua", "Xiahua" (b) Preparation of "Fanyin" (c) "Yao"

Feature Extraction. The fingering techniques closely relate to music expressiveness, including music emotion/mood aspects. Thus, we choose to transfer a feature set that focuses on speech emotion to present the audio characteristics of Guzheng fingering. This feature set comes from the INTERSPEECH 2009 Emotion Challenge [16]. The feature set covers 3 categories of features as the prosody features, the sound quality features, and the spectrum features, which are calculated from 12 statistical functions of 16 low-level descriptors (LLDs) and their first order difference, resulting into a $12 \times 16 \times 2 = 384$ dimensional feature vector.

The LLDs include zero-crossing-rate (ZCR), root mean square (RMS) energy, fundamental frequency (F0), harmonic-noise ratio (HNR), and first 12 Mel frequency cepstrum coefficients (MFCCs). ZCR refers to the ratio that the waveform goes across the horizontal axis, which is mainly used in speech analyzing to distinguish unvoiced and voiced sound. RMS energy presents the average level and changing tendency of the signal. F0 is the basic vibration frequency of a vibrating component, e.g., vocal cord for voice, strings for some instruments including Guzheng. It is perceived as pitch by human ears when hearing, and its changing rate and strength, which can be reflected by its statistics over a certain period of time, are strongly related to several fingering techniques, such as "Chanyin", "Shanghua", and "Xiahua". HNR is essentially used in evaluating voice quality. In analysis of Guzheng fingering, the harmonics come from the stable vibration of strings, and the noise part of signal comes from the hitting of the strings with the finger caps. Thus, the way and the frequency of hitting strings can be detected with this LLD. MFCCs are extracted on Mel scale, which follows the non-linear characteristics of the auditory characteristics of human ears, which can be expressed as:

$$Mel(f) = 2595 * \lg(1 + \frac{f}{700}). \tag{1}$$

The MFCCs can exhibit the timbre property of a sound, thus they are important for distinguishing different Guzheng fingering.

The above mentioned LLDs are typically calculated on frame level, and their statistics on a longer time scale can present the characteristics of a sound more adequately. The statistical functions in INTERSPEECH 2009 feature set are listed in Table 1.

Table 1. Statistical functions in INTERSPEECH 2009 feature set

Statistical functionals(12)
mean, standard deviation, kurtosis, skewness
maximum and minimum value, relative position, range
Linear regression: slope, skewness, mean square error (MSE)

The extraction of INTERSPEECH 2009 features set is processed with TUM's open-source openSMILE feature extractor [17], with the configuration "emo_IS09.conf".

Feature Analysis and Selection. Although the INTERSPEECH2009 feature set is only a small scaled feature set, the direct use of it in our work will still risk from the problem of "curse of dimensionality" [18] due to the limited number of audio samples in our dataset. We collected about 60 samples for each selected fingering, which means several samples on each string, that can be seen as approximately covering the basic playing of the corresponding fingering. The size of the INTERSPEECH 2009 feature set is 384-dimension, which is far more than 60 samples in a type of fingering. Moreover, there are a certain number of features in this feature set are not very suitable to present this plucked string instrument, such as ZCR related features. Thus, we added a feature selection before the automatic fingering recognition. Among the three main approaches of feature selection algorithms as Filter, Wrapper and Embedded algorithms [18,19], we choose a filter approach because we are using several different classifiers to analysis Guzheng fingering. Wrapper approaches and embedded approaches that rely on or embedded in the classifiers are not convenient in our case. A best 60-dimensional feature subset is selected in the classification.

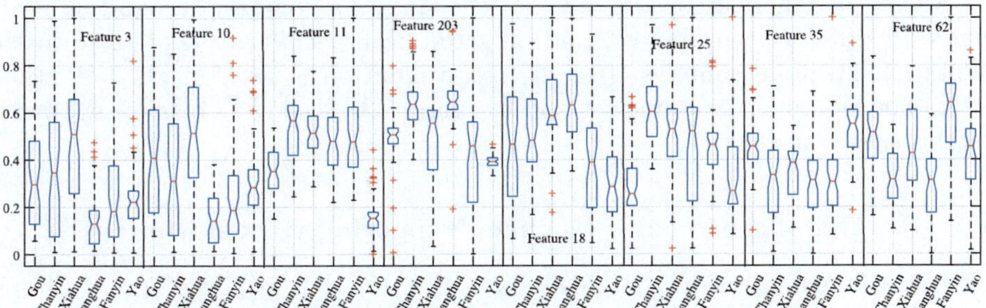

Fig. 3. Box-plot of selected features on the 6 fingerings

The box-plot to show the distribution range of several selected "good" features is displayed in Fig. 3. All the features are first normalized to the range of [0, 1] for convenient comparison. We can see from the comparison that the mean values of these features have obvious differences to provide ability in distinguishing, while their range are normally somehow overlapped with each other to prevent perfect recognizing. The content of the above shown features is listed in Table 2. Most of the effective features in distinguishing the Guzheng fingerings focus in RMS energy features and the first several MFCCs.

Table 2. List of selected feature

Index	Feature content	Index	Feature content
3	Range of RMS energy	18	Average of 1st MFCC
10	Standard deviation of RMS energy	25	Maximum of 2nd MFCC
11	Skewness of RMS energy	35	Skewness of 2nd MFCC
203	Skewness of 1st difference of RMS energy	62	Minimum of 5th MFCC

Choosing of Classifiers. For the automatic fingering recognition investigation on the small scaled dataset, traditional machine learning methods are used in this work instead of deep learning ones. Several different classifiers are adopted to avoid extremely high or low accuracies caused by inappropriate classifiers. The recognition algorithms are trained and tested on WEKA platform [20]. Rough tests were taken on a number of methods, and four classifiers with stable performances which come from 4 different WEKA categories are chosen for detailed recognition. The 4 chosen algorithms are: a minimum distance based method, k-nearest neighbors classifier (KNN), the corresponding method in WEKA is Lazy-IBK; a decision tree based classifier, Random Forest algorithm, the corresponding method in WEKA is Trees-RandomForest; a logistic regression based classifier, the corresponding method in WEKA is Functions-SimpleLogistic; and a support vector machine (SVM) based classifier, the corresponding method in WEKA is Functions-SMO.

All the classifiers are evaluated with the Guzheng fingering recognition problem with different parameters, and only the best results are kept. Due to the limitation of audio samples, 10-fold cross-validation is used in all evaluations instead of separating training set and testing set, to avoid the bias of separation on this small dataset.

3.2 Results of Automatic Guzheng Fingering Recognition

4 different machine learning algorithms as IBK (KNN), Random Forest, Simple Logistic, and SMO are evaluated in this subsection for automatic Guzheng fingering recognition. For each algorithm, the accuracies are calculated with feature subsets from 1 feature to 60 features as ranked in the filter feature selection.

The idea of the K nearest neighbor method (KNN) is to determine the category of an unknown sample x based on the most frequently occurring category of its K nearest neighborsby calculating the distance d(x,Y), where Y is the set of all known samples, and select K samples with the smallest distance as its neighbor sample. When $K = 1$, the K nearest neighbor method degenerates to the nearest neighbor method [21], which presents best performance of IBK in this work of Guzheng fingering recognition.

The Random Forest method is a combined classifier based on statistical learning theory, and an integrated learning method based on Bagging. Since sampling with replacement is used to construct training models with different data sets, the generalization ability of the model is usually stronger than that of a single model. This method leads to the overall best accuracy in our fingering recognition.

As a kind of generalized linear model, the Simple Logistic method does not perform so well as the other methods in this work, probably due to the fact that the fingering techniques are not typically linear separable with the selected features.

The widely used SVM approach, which aims to find an optimal classification hyperplane based on input data samples [22], is also evaluated. The actually used SVM is John Platt's SMO (sequential minimal optimization algorithm) [23]. The highest accuracies of the SMO classifier are obtained when the c parameter is set to 1.0.

The best accuracies of the fingering techniques from the 4 methods are illustrated in Fig. 4(a).

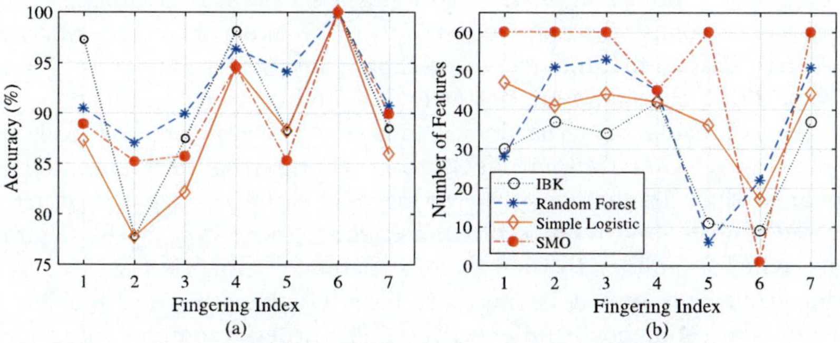

Fig. 4. Accuracy of the Guzheng fingering techniques recognition (a) Best accuracy (%) (b) Minimum number of features to get the best accuracy. The meaning of fingering index: 1-"Gou", 2-"Chanyin", 3-"Xiahua", 4-"Shanghua", 5-"Fanyin", 6-"Yao", 7-weighted average of the six fingerings

The overall accuracy of Guzheng fingering recognition on all 6 fingerings ranges from 85.96% (Simple Logistic) to 90.73% (Random Forest). For all 4 algorithms, the accuracies for each single fingering technique exhibits similar trends. The most basic fingering, "Gou", can be seen as a reference. There are 3 fingerings with F0 changing, where "Chanyin" gets the worse recognition, while "Shanghua" gets the best of the 3. The reason that the "Chanyin" cannot be very well recognized is that it can be seen as a combination of continuous "Shanghua" and "Xiahua" when with high strength, and it can be similar to "Gou" when it is very slight. "Fanyin" as a special fingering technique, the recognition rate is averagely a little lower than "Gou". This may partly come from the imperfect playing of some samples collected from exercising of students. The "Yao", which is played by hitting a certain string repeatedly and rapidly, is perfectly recognized by all 4 methods, because it possesses significant different properties than other fingerings.

The minimum numbers of features to achieve the corresponding best accuracies are illustrated in Fig. 4(b). The best recognized fingering, "Yao", also needs smallest number of features to get this perfect recognition, with an extreme case of only 1 feature using SMO. The fingerings with F0 changings generally need more features to be well recognized.

3.3 Detailed Analysis

In order to discover the similarity and difference between the selected fingerings in more details, we further analyzed the confusion patterns obtained from the above recognition results. The confusion matrices of the 4 classifiers when the accuracies are the highest are displayed in Fig. 5. The values are displayed with colors to be more intuitive to see.

First, "Yao", whose timbre characteristics are significantly different from all other fingerings, can be perfectly distinguished by all 4 classifiers. The other fingerings are recognized from around 75% to 97%, according to different classifiers, where IBK and Random Forest show good performance in "Shanghua", while the other 2 classifiers are better in "Chanyin".

For the confusion patterns, two common confusion patterns exist in all 4 classifiers. First, "Gou" is more likely to be confused as "Xiahua". This might be explained that the hitting of the string may cause the tension on the string to increase slightly, and may result in a slight higher pitch at the beginning of the tone, which is similar to the case of "Xiahua". "Gou" is also confused with other fingerings, because it is the most basic Guzheng playing technique, and all fingerings are somehow similar with it. The other common confusion is that "Fanyin" is more likely to be confused as "Chanyin". One possible explanation is that the most obvious character of "Chanyin" is the continuous changing of F0, while the F0 of "Fanyin" is usually very weak and tends to be detected with very bad accuracy. The 3 fingerings with F0 changing also tend to be confused. For IBK and Random Forest, "Shanghua" is very well recognized with accuracies nearly 97%, but "Chanyin" and "Xiahua" are confused with each other with relatively high rates; for Simple Logistic and SMO, "Shanghua" and "Xiahua", which both have monotonic changing of F0, are highly confused with each other.

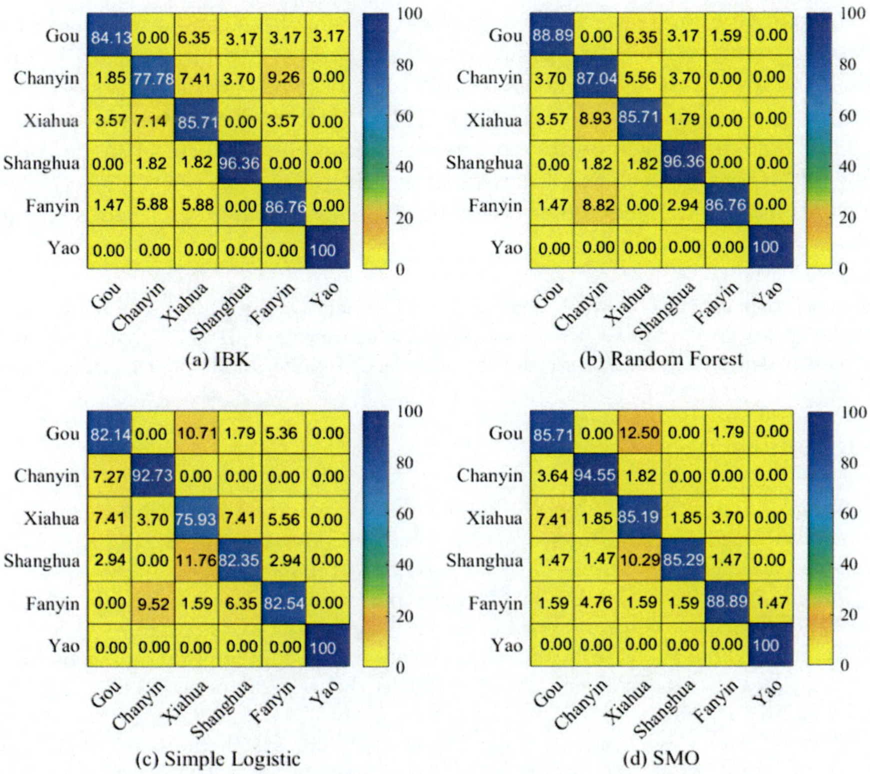

Fig. 5. Confusion matrice in Guzheng fingering recognition

The confusion patterns among different classifiers have some common tendencies. This proved that the Guzheng fingerings can be effectively distinguished with machine learning approaches, and that the feature set adopted from speech emotion analysis also works well on Guzheng, which is a kind of plucked string instrument. There are also some different confusion patterns according to the mechanism of classifiers. Due to the limited scale and source of collected Guzheng audio samples, advantages of the selected classifiers are not fully presented in this work. For example, larger data scale will be better to further use the random selection property of Random Forest algorithm.

4 Conclusion

Six typical fingering techniques of the Guzheng playing are analyzed with automatic recognition in this paper. The audio samples partly come from CCMusic by the China Conservatory of Music and partly come from personal recording samples in excersicing. A feature set is adopted from speech emotion analysis for the analyzing of Guzheng fingerings. Four classifier algorithms were selected for comparative analysis. The 6 selected Guzheng fingerings are proved to be separable with the adopted feature set, with the best overall accuracy obtained from Random Forest algorithm as 90.73%. The best recognized fingering, "Yao", is perfectly recognized with accuracy of 100% with all 4 classifiers.

The Guzheng fingerings are currently evaluated in the form of single tones in this work. In our future work, we will extend the automatic fingering recognition into continuous playing of whole Guzheng music pieces on much larger scale data, to make it more practical.

Acknowledgment. This work was supported in part by the National Natural Science Foundation of China under Project 61906128 and Project 61802272, in part by the National Natural Science Foundation of Jiangsu Province under project BK20180834.

References

1. Wang, X.: A discussion on the application of traditional rhythm techniques in modern Guzheng performance. Int. J. Intell. Inf. Manage. Sci. **8**(2) (2019). http://en.cnki.com.cn/Article_en/CJFDTotal-YSLL201703036.htm
2. Dai, X.: Analysis of timbre in Guzheng performance. Song Yellow River **13**, 31 (2019). https://doi.org/10.3969/j.issn.1810-2980.2019.13.024
3. Yang, Y., Yang, P.: A Probe into the development of contemporary guzheng playing techniques. Northern music **16**, 7–8 (2019). https://doi.org/10.3969/j.issn.1002-767X.2019.16.008
4. Chen, C.: The evolution of fingering of Guzheng in Zhejiang in 1960s. Songs Bimonthly **3**, 58–61 (2020). https://doi.org/10.3969/j.issn.1007-4910.2020.03.016
5. Wang, Z., Cao, Y.: An on-line algorithm for music-to-score alignment of Guzheng performance. In: 2018 IEEE 23rd International Conference on Digital Signal Processing (DSP), Shanghai, China (2018). https://doi.org/10.1109/ICDSP.2018.8631834
6. Liang S., Su, A., Lin C.: A new recurrent-network-based music synthesis method for Chinese plucked-string instruments - Pipa and Qin. In: International Joint Conference on Neural Networks, DC, USA (1999). https://doi.org/10.1109/IJCNN.1999.833478
7. Chen, Y., Huang, C.: Sound synthesis of the pipa based on computed timbre analysis and physical modeling. IEEE J. Sel. Topics Signal Process **5**(6), 1170–1179 (2011). https://doi.org/10.1109/JSTSP.2011.2162816
8. Wicaksana, H., Hartono, S., Wei, F.S.: Recognition of musical instruments. In: Proceedings of the IEEE Asia Pacific Conference on Circuits and Systems, pp. 327–330. Nanyang Technological University, Singapore(2006). https://doi.org/10.1109/APCCAS.2006.342417
9. Lin, Y.: Research on musical instrument recognition based on acoustic features. South China University of Technology (2012)
10. Wang, F.: Research on instrument recognition method based on tone analysis and deep learning. Jiangnan University (2018)
11. Ren, T.: Research on recognition of several typical musical instruments based on acoustic features. Northeast Forestry University (2018)
12. Wang, Q.: Tone recognition of western musical instruments. Shan Dong University (2015). http://d.wanfangdata.com.cn/thesis/Y1867745

13. Jeyalakshmi, C., Murugeshwari, B., Karthick, M.: HMM and K-NN based automatic musical instrument recognition. In: 2018 2nd International Conference on I-SMAC (IoT in Social, Mobile, Analytics and Cloud) (I-SMAC) I-SMAC (IoT in Social, Mobile, Analytics and Cloud) (I-SMAC), Palladam, India, pp. 350–355 (2018). https://doi.org/10.1109/I-SMAC.2018.8653725
14. Zhao, Y.: Technical analysis of Guzheng Finger shaking and tone formation. HuangZhong (J. Wuhan Conservatory Music) **1**, 117–120 (2006). https://doi.org/10.3969/j.issn.1003-7721.2006.01.017
15. Li Z., Yu S., Xiao C.: CCMusic database: construction of Chinese music database for MIR research. J. Fudan Univ. (Nat. Sci.) **3**, 351–357 (2019). https://doi.org/10.15943/j.cnki.fdxb-jns.2019.03.007
16. Eyben, F., Schuller, B.: openSMILE:). ACM SIG Multimedia Rec. **6**(4), 4–13 (2015). https://doi.org/10.1145/2729095.2729097
17. Eyben, F., Schuller, B.: openSMILE – the Munich versatile and fast open-source audio feature extractor. In: Proceedings of the 9th ACM International Conference on Multimedia, pp. 1459–1462 (2010). https://doi.org/10.1145/1873951.1874246
18. Kojadinovic, I., Wottka, T.: Comparison between a filter and a wrapper approach to variable (2001). https://core.ac.uk/display/24301854
19. Sebban, M., Nock, R.: A hybrid filter/wrapper approach of feature selection using information theory. Pattern Recognit. **35**(4), 835–846 (2002). https://doi.org/10.1016/s0031-3203(01)00084-x
20. Hall, M., Frank, E., Holmes, G.: The WEKA data mining software: an update. ACM SIGKDD Explor. News. **11**, 10–18 (2009). https://doi.org/10.1145/1656274.1656278
21. Xin, Y.: Research and implementation of handwriting recognition system based on k-neighbor algorithm. Electron. Design Eng. **26**(7), 27–30 (2018). https://doi.org/10.3969/j.issn.1674-6236.2018.07.007
22. Liu, F., Wang, S.: Summary of support vector machine model and application. Comput. Syst. Appl. **27**(4), 1–9 (2018). https://doi.org/10.15888/j.cnki.csa.006273
23. Wang, G.: Research on theory and algorithm for support vector machine classifier. Beijing Univ. Posts Telecommun. (2007). http://qikan.cqvip.com/Qikan/Article/Detail?id=36534620

Check for updates

MusicTM-Dataset for Joint Representation Learning Among Sheet Music, Lyrics, and Musical Audio

Donghuo Zeng(✉), Yi Yu, and Keizo Oyama

National Institute of Informatics, SOKENDAI, Tokyo, Japan
{zengdonghuo,yiyi,oyama}@nii.ac.jp

Abstract. This work present a music dataset named MusicTM-Dataset, which is utilized in improving the representation learning ability of different types of cross-modal retrieval (CMR). Little large music dataset including three modalities is available for learning representations for CMR. To collect a music dataset, we expand the original musical notation to synthesize audio and generated sheet-music image, and build musical notation based sheet-music image, audio clip and syllable-denotation text as fine-grained alignment, such that the MusicTM-Dataset can be exploited to receive shared representation for multi-modal data points. The MusicTM-Dataset presents 3 kinds of modalities, which consists of the image of sheet-music, the text of lyrics and synthesized audio, their representations are extracted by some advanced models. In this paper, we introduce the background of music dataset and express the process of our data collection. Based on our dataset, we achieve some basic methods for CMR tasks. The MusicTM-Dataset are accessible in https://github.com/dddzeng/MusicTM-Dataset.

Keywords: MusicTM-Dataset · MIR · Canonical correlation analysis

1 Introduction

Music data is getting readily accessible in digital form online, which brings difficult to manage the music from a large amount of personal collection. It highly relies on the music information retrieval to retrieve the right data information for users. In recent years, machine learning or deep learning based methods has become increasing prevailing in music information retrieval [1–8] and has played an essential role in MIR.

This paper concentrates on content music MIR by learning semantic concepts across different music modalities for MIR, as shown in the Fig. 1. For instance, when we play music audio, we want to find what is the corresponding sheet music and which lyrics is correct, by learning two kinds of relationship in audio-sheet music and audio-lyrics. Such kinds of relationship obtained from content-based representation by learning the alignment across two modalities in the shared

X. Shao et al. (Eds.): CSMT 2020, LNEE 761, pp. 78–89, 2021.
https://doi.org/10.1007/978-981-16-1649-5_7

latent subspace without introducing any users' information. The unsupervised representation learning method ensures the system can allow users to find the right music data modalities with the other data modalities as query.

The major challenge of unsupervised representation learning for different music modalities is the modality gap. Representation learning for two music data modalities such as audio-lyrics [9–11], audio-sheet music [12,13], have become increasingly in the CMR task to bridge the modality gap. In previous works, classic CCA and CCA-variant methods [14,15] are popular in representation learning between two music data modalities, through finding linear or nonlinear transformation to optimize the correlation between two data modalities in the shared latent subspace. With the success of Deep Neural Network (DNN) in representation learning, DNN is also helpful for learning joint representation for cross-modal tasks [16], for example, attention network [12] applies a soft-attention mechanism for the audio branch to learn the relationship between sheet music and audio, which solves the problem that the music recordings easily brings about the global and local time deviations.

However, representation learning for two modalities is still not enough to achieve the music information retrieval, when we apply one data modality as query to retrieve other two different data modalities. The existing dataset normally applied in learning correlation between two modalities in a shared space. The paper [13] collect a dataset contains an alignment between sheet music and music audio, which explores music audio to find the corresponding sheet music snippets. [17] apply a lyrics and audio paired dataset to align lyrics to audio. In this paper, we collect a new music dataset including three music data modalities. In particular, sheet music and audio are generated from music notes by music generation tools, the syllable-level lyrics and music notes are fine-grained alignment. Three major contributions of this paper have achieved in the following aspects: 1) we collect a fine-grained alignment across three music data modalities, which is useful for representation learning methods to obtain high-level feature for music CMR tasks. 2) we release experimental results of some baselines such as CCA and Generalized CCA on our MusicTM-Dataset. 3) The performance of Generalized CCA surpasses the CCA on audio-sheet music CMR task, which shows that the mapping all the three data modalities into a shared latent subspace can be better than mapping them into two shared latent subspace for audio-sheet music cross-modal retrieval.

The rest parts are arranged as follows. Some existing related works show in Sect. 2. In Sect. 3, we explain the detail of our data collection, feature representations and the metrics we applied on our experiment in Sect. 4. Section 5 makes a conclusion of the whole paper.

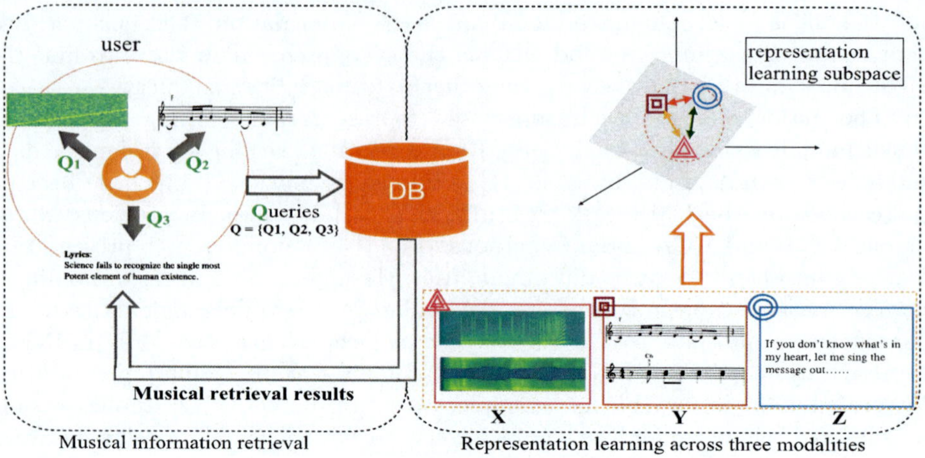

Fig. 1. The framework of representation learning for music information retrieval.

2 Related Works

2.1 Audio and Lyrics

Recently, the study of automatic audio-lyrics alignment techniques is getting trendy. The aim of the topic is to estimate the relation between audio and lyrics, such as temporal relation [18], deep sequential correlation [19]. [17] establishes audio-lyrics alignment based on a hidden Markov model speech recognizer, in particular, the lyrics input is to create a language model and apply the Viterbi method to link the audio and the lyrics. Synchronizing lyrics information with an audio recording is an important music application. [20] presents an approach for audio-lyric alignment by matching the vocal track and the synthesized speech.

2.2 Sheet Music and Audio

The popular problem of correlation learning between sheet music and audio is to establish the relevant linking structures between them. In [21], it aims to establish linking the regions of sheet music to the corresponding piece in an audio of the same clip. [22] bring forwards an multi-modal convolutional neural network, by taking an audio snippet as input to find the relevant pixel area in sheet music image. However, the global and local tempo deviations in music recordings will influence the performance of the retrieval system in the temporal context. To address that, [23] introduces an additional soft-attention mechanism on audio modality. Instead of correlation learning with high-level representations, [13] matches music audio to sheet music directly, the proposed method learns shared embedding space for short snippet of music audio and the corresponding piece in sheet music.

2.3 Lyrics and Sheet Music

Learning the correlation between lyrics and sheet music is a challenging research issue, which requires to learning latent relationship with high-level representations. The automatic composition techniques are considerable for upgrading music applications. [24] proposed a novel deep generative model LSTM-GAN to learn the correlation in lyrics and melody for generation task. Similarly, [25] presents an approach that is used to generate music song from a Japanese lyrics. [26] introduces a novel language model that can generate lyrics from a given sheet music. [27] presents an better query in using lyrics and melody, which take advantage of extra lyrics information by linking the scores from pitch-based lyrics and melody recognition. Accept that, "singing voice," which is for generating singing voice has been drawing attention in the last years, [28] explores a novel model that the singing voice generation with no consideration of pre-assigned melody and lyrics.

3 Dataset and Metrics

This section presents the motivation and contribution of our data collection. Moreover, also the process of dataset collection applied in our experiments and the data feature extraction are discussed. In the end, we show all the evaluation metrics applied to leverage our models.

3.1 Dataset Collection

Figure 2 shows a few examples of MusicTM-Dataset we applied, including the spectrum of music audio with Librosa library[1], word-level lyrics, and sheet music with Lilypond technique[2].

The available music dataset with three modalities, which can be applied in music information retrieval based on the high-level semantic features is rarely reported. We try to learn aligned representation for sheet music images, music audio, and lyrics because they frequently appear in the music data collection. We follow the work [24] to collect our music dataset by extending two modalities (lyrics and music notes) to three modalities: sheet music, audio, and lyrics.

In [24] presents a music dataset that a music is represented by lyrics and music notes. The lyrics is parsed as syllable-level collection, such as the lyrics: 'Listen to the rhythm of fall ...' will parse as 'Lis ten to the rhy thm of fall'. A music note is a ternary structure that includes three attributions: pitch, duration, and rest. The pitch is a frequency-related scale of sounds, for example, piano keys MIDI number ranges from 21 to 108, each MIDI number corresponds to a pitch number, such as MIDI number '76' represents pitch number 'E5'. Duration in music notes denotes the time of the pitch, for example, a pitch number 'E5' with its duration 1.0, means this music note will last 0.5 s in the playing. The rest of

[1] https://librosa.org/doc/latest/index.html.
[2] http://lilypond.org/.

Fig. 2. Examples of three data modalities in MusicTM-Dataset.

the pitch is the intervals of silence between two adjacent music notes, which share the same unit with duration. The dataset used for the melody generation from lyrics, to consider the time-sequence information in the pairs, the syllable-level lyrics and music notes are aligned by pairing a syllable and a note.

The initial pre-processing for our dataset is to get the beginning of music notes and corresponding syllables. In our MusicTM-Dataset collection, we adopted the same method to get the first 20 notes as a sample and ensure the syllable-level lyrics corresponding can be kept. Moreover, we removed the samples if existing the rest attributes of the note are longer than 8 (about four seconds).

Music audio and sheet music are separately created from music notes that matches our purpose of musical multimodal building. We use syllable-level lyrics and notes to create the pairs of sheet and audio by some high-quality technologies. All the music data modalities contain temporal structure information, which motivates us to establish fine-grained alignment across different modalities, as seen in Fig. 3. In detail, the syllable of lyrics, the audio snippet, and sheet music fragment generated from music notes are aligned.

Music audio is also music sound transmitted in signal form. We add piano instrument in the music channel to create new midi files, and synthesize audios with TiMidity++ tool[3].

Sheet music is created by music note with Lilypond tools. Lilypond is a compiled system that runs on a text file describing the music. The text file may contain music notes and lyrics. The output of Lilypond is sheet music which can be viewed as an image. Lilypond is like a programming language system, music notes are encoded with letters and numbers, and commands are entered with backslashes. It can combine melody with lyrics by adding the "\addlyrics" command. In our MusicTM-Dataset, sheet music (visual format) for one note and entire sheet music (visual format) for 20 notes are created respectively. Accordingly, each song has single note-level and sequential note-level (sheet fragment) visual formats.

[3] http://timidity.sourceforge.net/.

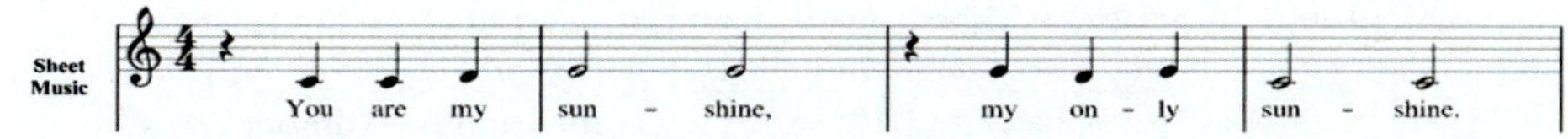

Lyrics	You	are	my	sun	shine	my	on	ly	sun	shine
Music note	D5	G3	A2	B1	B1	B1	A2	B5	G3	G3
Duration	0.5	1	0.5	1	0.5	1	1	0.5	2.5	1
Reset	0	0	0	1	1	0	1	0	1	1

Fig. 3. An example of fine-grained alignment across three modalities.

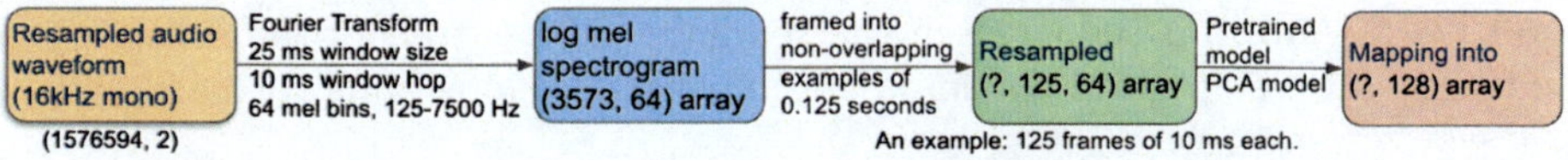

Fig. 4. The audio feature extraction process with vggish model

3.2 Feature Extraction

This section will explain the feature extraction for music multimodal data.

Audio Feature Extraction. Generally, audio signal is used for audio feature extraction, which plays the main role in speech processing [29,30], music genre classification [31], and so on. Here, we present a typical model for audio feature extraction, the supervised trained model Vggish. The detailed process of feature extraction can be seen in Fig. 4. Firstly, we resample audio waveform 16 kHz mono, then calculate a spectrogram. Secondly, in order to obtain a stable log mel spectrogram, it is computed by exploring log. Finally, resampling the feature into (125,64) format, then applying pre-trained model to extract feature and use PCA model to map it into 128-dimensional.

Sheet Music Feature Extraction. Different from other image feature extraction, our feature extraction of sheet music image tries to catch pitches and the segments. In this paper, our information extraction of sheet music has two levels, pitch detection, and semantic segments. We apply the ASMCMR [32] model trained in audio-sheet retrieval tasks, which learns the correlation between audio clips and corresponding sheet snippets. In our work, the shape of extracted note-level feature and sheet snippet-level features are (100, 32) and (32,) respectively.

Table 1. Statistics of MusicTM-Dataset applied in our experiments

Modality	Feature extractor	Dimension	Number
Audio	Vggish	(20, 128)	14,454
Lyrics	Skip-gram	(20, 20)	14,454
Sheet music	Lilypond&ASMCMR	(20, 100, 32)	14,454

Lyrics Feature Extraction. We follow [24] to keep the alignment between syllable and note by representing lyrics in the form of syllable and word level. The syllable-level feature extracted with the syllable skip-gram model, the word-level feature extracted with the word skip-gram model used in [24]. These two pre-trained skip-gram models are trained on all the lyrics data, which applied in a regression task with SGD optimization. The input of syllable-level skip-gram model is a sequence of syllables in a sentence, while the input of word-level model is a word unit sequence in the sentence. The output of the syllable-level and word-level skip-gram model is 20-dimensional embedding for each syllable and word, respectively.

The overall statistics of our music data are shown in Table 1. We divided the dataset into 3 parts as training, validation, and testing set by 70%, 15%, and 15%. The number of training, validation, and testing set are 13,535, 2800, and 2800 respectively.

3.3 Evaluation Metric

To evaluate some baselines on our dataset, we apply some standard evaluation from the work [33] for unsupervised learning based cross-modal retrieval. R@K (Recall at K, here we set K as 1, 5, and 10) is to compute correct rate that is the percentage of retrieved items corresponding to the query in the top-K of rank list. Fox instance, R@1 calculate the percentage of sample appear in the first item of retrieved list. In order to further evaluate our collected dataset with some baselines, we also apply the Median Rank and Mean Rank to compute the mean and median rank of all the correct results.

4 Experiments

Table 2. The performance of multimodal information retrieval on MusicTM-Dataset.

Methods	R@1	R@5	R@10	MedR	MeanR
audio2lyrics retrieval					
Random rank [34]	0.028	0.055	0.076	7312.0	7257.2
CCA [35]	0.306	0.350	0.353	423.0	639.4
GCCA [36]	0.040	0.074	0.093	770.0	881.1
lyrics2audio retrieval					
Random rank	0.027	0.055	0.076	7316.0	7257.3
CCA	0.304	0.349	0.354	427.0	639.3
GCCA	0.039	0.078	0.095	774.0	881.6
sheet music2lyrics retrieval					
Random rank	0.027	0.055	0.075	7311.0	7257.3
CCA	0.093	0.172	0.203	524.0	708.7
GCCA	0.089	0.0142	0.167	573.0	770.5
lyrics2sheet music retrieval					
Random rank	0.027	0.055	0.077	7313.0	7257.4
CCA	0.093	0.168	0.198	522.0	709.0
GCCA	0.098	0.014	0.168	578.0	769.8
audio2sheet music retrieval					
Random rank	0.028	5.57	7.50	7310.0	7257.2
CCA	0.303	0.349	0.353	341.0	596.5
GCCA	0.358	0.403	0.414	271.0	382.8
sheet music2audio retrieval					
Random rank	0.026	0.055	0.075	7310.0	7257.4
CCA	0.300	0.350	0.354	332.0	596.1
GCCA	0.362	0.407	0.415	271.0	381.3

4.1 Baselines

CCA can be seen as the method that aims at finding linear transforms for two sets of variables in order to optimize the relation between the projections of the variable sets into a shared latent subspace. Consider two variables from two data modalities $X \in R^{D_x}$ and $Y \in R^{D_y}$ with zero mean and the two paired data sets $S_x = \{x_1, x_2, ..., x_n\}$ and $S_y = \{y_1, y_2, ..., y_n\}$. $W_x \in R^{D_x}$ and $W_y \in R^{D_y}$ as the directions that linearly map the two set into a shared latent subspace, such that the relation between the projection of S_x and S_y on W_x and W_y is optimized.

$$\rho = \arg\max_{(W_x, W_y)} \frac{W_x^T \Sigma_{xy} W_y}{\sqrt{W_x^T \Sigma_{xx} W_x \cdot W_y^T \Sigma_{yy} W_y}} \tag{1}$$

where ρ is the correlation, Σ_{xx} and Σ_{yy} denote the variance–covariance matrix of S_x, S_y, respectively and Σ_{xy} represents the cross-covariance matrix.

Generalized CCA [36] can be viewed as an extension method of CCA, which aims to solve the limitation on the number of data modalities. The objective function in Eq. 2, which focuses on finding a shared representation G for K different data modalities.

$$minimize_{(W_k,G)} = \sum_{k=1}^{K} ||G - W_k^T X_k||_F^2 \quad (2)$$

where K is the size of data points, and X_k is a matrix for k^{th} data modality. Similar to CCA, GCCA is to find linear transformation for different data modalities to optimize the correlation within them.

4.2 Results

In Table 2, when learning the correlation between two data modalities with CCA method, the correlation of audio-lyrics and audio-sheet music can get more than 30% of R@1, which illustrates the dataset can be learned for cross-modal retrieval task. Specifically, in comparison with CCA and RANDOM, GCCA will have a big improvement in the performance of audio-sheet music cross-modal retrieval. In detail, compared with CCA method, 5.46%, 5.39%, 6.06%, 70, and 213.68 improved in R@1, R@5, R@10, MedR, and MeanR for music audio as the query to retrieve the correct sheet music; 6.16%, 5.65%, 6.1%, 61, and 214.8 improved in R@1, R@5, R@10, MedR, and MeanR for sheet music as the query to retrieve the correct music audio. However, GCCA will decrease the performance of audio-lyrics cross-modal retrieval and achieve a similar performance of sheet music-lyrics cross-modal retrieval.

The results show that the learned representation with GCCA for sheet image, lyrics, and music audio can raise the relation of sheet music and music audio. However, such representations drop the correlation between music audio and lyrics and their correlation between sheet music image and lyrics will almost stay the same as CCA method, which learns the representation in the shared subspace without involving lyrics data. The results prove our hypothesis can be accepted that the sheet music and music audio are created by music notes, so the correlation between audio and sheet music will be close. The lyrics and music note from original dataset exist alignment between each other, the correlation between the two can be learned. In this case, the correlation between audio and lyrics reflects the correlation between audio and music note, however, the correlation between sheet music and lyrics seems hard to learn.

In visualization of the the position of sheet music, lyrics, and music audio in CCA and GCCA subspace, as shown in Fig. 5. GCCA seems to pull audio and sheet music while pushing the audio and lyrics compared with the CCA subspace. This motivates us to propose a new advanced model that can improve three couples of cross-modal retrieval tasks in a shared latent subspace as the GCCA subspace achievement in the future.

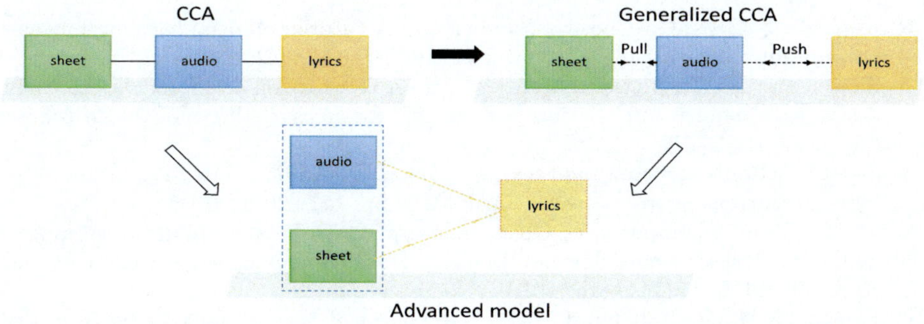

Fig. 5. The general paradigm of MusicTM-Dataset with two different models (CCA, GCCA)

5 Conclusion

This paper presents a MusicTM-Dataset that consists of three different data modalities and there is fine-grained alignment across the modalities. The dataset can be easily extended to different researches, we report the performance of some baselines on our MusicTM-Dataset, which allows the results of the following research to be compared. Instead of applying CCA to learn shared latent subspace for every two modalities, GCCA learns the correlation of three modalities in one shared latent subspace. The performance of audio-sheet music can be improved and the performance of audio-lyrics cross-modal retrieval is quilt similar but the performance of lyrics-sheet music cross-modal retrieval will be decreased. In theory, we want to develop a new architecture that will improve the performance of multimodal information retrieval across different modalities.

Acknowledgements. The JSPS Grant for SR financed this work, which is under Grant No. 19K11987.

References

1. Eyben, F., Böck, S., Schuller, B., Graves, A.: Universal onset detection with bidirectional long-short term memory neural networks. In: Proceedings of the 11th International Society for Music Information Retrieval Conference, ISMIR, Utrecht, The Netherlands, pp. 589–594 (2010)
2. Hamel, P., Eck, D.: Learning features from music audio with deep belief networks. In: ISMIR, Utrecht, The Netherlands, vol. 10, pp. 339–344 (2010)
3. Zhou, X., Lerch, A.: Chord detection using deep learning. In: Proceedings of the 16th ISMIR Conference, vol. 53, p. 152 (2015)
4. Böck, S., Krebs, F., Widmer, G.: Accurate tempo estimation based on recurrent neural networks and resonating comb filters. In: ISMIR, pp. 625–631 (2015)
5. Grill, T., Schluter, J.: Music boundary detection using neural networks on spectrograms and self-similarity lag matrices. In: 2015 23rd European Signal Processing Conference (EUSIPCO), pp. 1296–1300. IEEE (2015)

6. Choi, K., Fazekas, G., Cho, K., Sandler, M.: A tutorial on deep learning for music information retrieval. arXiv preprint arXiv:1709.04396 (2017)
7. Siedenburg, K., Fujinaga, I., McAdams, S.: A comparison of approaches to timbre descriptors in music information retrieval and music psychology. J. New Music Res. **45**(1), 27–41 (2016)
8. Sigtia, S., Boulanger-Lewandowski, N., Dixon, S.: Audio chord recognition with a hybrid recurrent neural network. In: ISMIR, pp. 127–133 (2015)
9. Yu, Y., Tang, S., Raposo, F., Chen, L.: Deep cross-modal correlation learning for audio and lyrics in music retrieval. ACM Trans. Multimedia Comput. Commun. Appl. (TOMM) **15**(1), 1–16 (2019)
10. Kruspe, A.M., Fraunhofer, I.D.M.T.: Retrieval of textual song lyrics from sung inputs. In: INTERSPEECH, pp. 2140–2144 (2016)
11. Kruspe, A.M., Goto, M.: Retrieval of song lyrics from sung queries. In: 2018 IEEE International Conference on Acoustics, Speech and Signal Processing (ICASSP), pp. 111–115. IEEE (2018)
12. Balke, S., Dorfer, M., Carvalho, L., Arzt, A., Widmer, G.: Learning soft-attention models for tempo-invariant audio-sheet music retrieval. arXiv preprint arXiv:1906.10996 (2019)
13. Matthias, D., Hajič Jr., J., Arzt, A., Frostel, H., Widmer, G.: Transactions of the International Society for Music Information Retrieval **1**(1) (2018)
14. Dorfer, M., Arzt, A., Widmer, G.: Towards end-to-end audio-sheet-music retrieval. arXiv preprint arXiv:1612.05070 (2016)
15. Dorfer, M., Schlüter, J., Vall, A., Korzeniowski, F., Widmer, G.: End-to-end cross-modality retrieval with CCA projections and pairwise ranking loss. Int. J. Multimedia Inf. Retrieval **7**(2), 117–128 (2018)
16. Yu, Y., Tang, S., Aizawa, K., Aizawa, A.: Category-based deep CCA for fine-grained venue discovery from multimodal data. IEEE Trans. Neural Netw. Learn. Syst. **30**(4), 1250–1258 (2018)
17. Mauch, M., Fujihara, H., Goto, M.: Integrating additional chord information into hmm-based lyrics-to-audio alignment. IEEE Trans. Audio Speech Lang. Process. **20**(1), 200–210 (2011)
18. Fujihara, H., Goto, M.: Lyrics-to-audio alignment and its application. In: Müller, M., Goto, M., Schedl, M. (eds.) Multimodal Music Processing. Dagstuhl Follow-Ups, vol. 3, pp. 23–36. Schloss Dagstuhl - Leibniz-Zentrum fuer Informatik, Germany (2012)
19. Yu, Y., Tang, S., Raposo, F., Chen, L.: Deep cross-modal correlation learning for audio and lyrics in music retrieval. CoRR, abs/1711.08976 (2017)
20. Lee, S.W., Scott, J.: Word level lyrics-audio synchronization using separated vocals. In: 2017 IEEE International Conference on Acoustics, Speech and Signal Processing, ICASSP 2017, New Orleans, LA, USA, 5–9 March 2017, pp. 646–650. IEEE (2017)
21. Thomas, V., Fremerey, C., Müller, M., Clausen, M.: Linking sheet music and audio - challenges and new approaches. In: Müller, M., Goto, M., Schedl, M. (eds.) Multimodal Music Processing. Dagstuhl Follow-Ups, vol. 3, pp. 1–22. Schloss Dagstuhl - Leibniz-Zentrum fuer Informatik, Germany (2012)
22. Dorfer, M., Arzt, A., Widmer, G.: Towards score following in sheet music images. In: Mandel, M.I., Devaney, J., Turnbull, D., Tzanetakis, G. (eds.) Proceedings of the 17th International Society for Music Information Retrieval Conference, ISMIR 2016, New York City, United States, 7–11 August 2016, pp. 789–795 (2016)

23. Balke, S., Dorfer, M., Carvalho, L., Arzt, A., Widmer, G.: Learning soft-attention models for tempo-invariant audio-sheet music retrieval. In: Flexer, A., Peeters, G., Urbano, J., Volk, A. (eds.) Proceedings of the 20th International Society for Music Information Retrieval Conference, ISMIR 2019, Delft, The Netherlands, 4–8 November 2019, pp. 216–222 (2019)
24. Yu, Y., Canales, S.: Conditional LSTM-GAN for melody generation from lyrics. arXiv preprint arXiv:1908.05551 (2019)
25. Fukayama, S., Nakatsuma, K., Sako, S., Nishimoto, T., Sagayama, S.: Automatic song composition from the lyrics exploiting prosody of the Japanese language. In: Proceedings of the 7th Sound and Music Computing Conference (SMC), pp. 299–302 (2010)
26. Watanabe, K., Matsubayashi, Y., Fukayama, S., Goto, M., Inui, K., Nakano, T.: A melody-conditioned lyrics language model. In: Walker, M.A., Ji, H., Stent, A. (eds.) Proceedings of the 2018 Conference of the North American Chapter of the Association for Computational Linguistics: Human Language Technologies, NAACL-HLT 2018, New Orleans, Louisiana, USA, 1–6 June 2018, vol. 1 (Long Papers), pp. 163–172. Association for Computational Linguistics (2018)
27. Wang, C.-C., Roger Jang, J.-S., Wang, W.: An improved query by singing/humming system using melody and lyrics information. In: Stephen Downie, J., Veltkamp, R.C. (eds.) Proceedings of the 11th International Society for Music Information Retrieval Conference, ISMIR 2010, Utrecht, Netherlands, 9–13 August 2010, pp. 45–50. International Society for Music Information Retrieval (2010)
28. Liu, J.-Y., Chen, Y.-H., Yeh, Y.-C., Yang, Y.-H.: Score and lyrics-free singing voice generation. arXiv preprint arXiv:1912.11747 (2019)
29. Watanabe, S. Hori, T., Karita, S., Hayashi, T., Nishitoba, J., Unno, Y., Soplin, N.E.Y., Heymann, J., Wiesner, M., Chen, N., et al.: ESPnet: end-to-end speech processing toolkit. arXiv preprint arXiv:1804.00015 (2018)
30. Lambert, C., Kormos, J., Minn, D.: Task repetition and second language speech processing. Stud. Second Lang. Acquisition **39**(1), 167–196 (2017)
31. Kobayashi, T., Kubota, A., Suzuki, Y.: Audio feature extraction based on subband signal correlations for music genre classification. In: 2018 IEEE International Symposium on Multimedia (ISM), pp. 180–181. IEEE (2018)
32. Dorfer, M., Arzt, A., Widmer, G.: Learning audio-sheet music correspondences for score identification and offline alignment. In: Proceedings of the 18th International Society for Music Information Retrieval Conference, ISMIR 2017, Suzhou, China, 23–27 October 2017, pp. 115–122 (2017)
33. Otani, M., Nakashima, Y., Rahtu, E., Heikkilä, J., Yokoya, N.: Learning joint representations of videos and sentences with web image search. In: Computer Vision - ECCV 2016 Workshops - Amsterdam, The Netherlands, 8–10 and 15–16 October 2016, Proceedings, Part I, pp. 651–667 (2016)
34. Zeng, D., Yu, Y., Oyama, K.: Unsupervised generative adversarial alignment representation for sheet music, audio and lyrics. arXiv preprint arXiv:2007.14856 (2020)
35. Hardoon, D.R., Szedmák, S., Shawe-Taylor, J.: Canonical correlation analysis: an overview with application to learning methods. Neural Comput. **16**(12), 2639–2664 (2004)
36. Tenenhaus, A., Tenenhaus, M.: Regularized generalized canonical correlation analysis. Psychometrika **76**(2), 257 (2011)

General Audio Signal Processing

Adversarial Domain Adaptation for Open Set Acoustic Scene Classification

Chunxia Ren[1] and Shengchen Li[2(✉)]

[1] Beijing University of Posts and Telecommunications, No. 10 Xitucheng Road, Haidian District, Beijing, China
chunxiaren@bupt.edu.cn

[2] Department of Intelligent Science, School of Advanced Technology, Xi'an Jiaotong-Liverpool University, 111 Ren'ai Road, Suzhou Industrial Park, Suzhou 215123, Jiangsu Province, P. R. China
shengchen.li@xjtlu.edu.cn

Abstract. Many algorithms classify acoustic scenes with predefined acoustic scenes categories but few addresses identifying acoustic scenes that are not predefined (usually referred as "unknown acoustic scenes"), which is known as "open set" problem for acoustic scene classification. Traditional methods generally use a "one-size-fits-all" threshold to make a second judgment on the output of trained model. The boundary between known and unknown scenes cannot be learned. To enable this boundary to be programmed, this paper proposes a novel method to introduce adversarial domain adaptation into the open set acoustic scene classification. In this method, known scenes are classified through the adaptation of target domain and source domain, and unknown scenes are distinguished by adversarial training with the help of preset pseudo-threshold. Not only the discrimination between unknown classes and known classes can be learned during the adversarial training process, but the overall performance of the open set acoustic scene classification algorithm is also improved. The proposed system achieves better performance compared with the baseline of open set acoustic scene detection in Detection and Classification on Acoustic Scenes and Events challenge 2019.

Keywords: Open set · Acoustic scene classification · Adversarial domain adaptation · Pseudo-threshold

1 Introduction

The task of identifying the acoustic scene according to characteristics of the audio is called acoustic scene classification (ASC) [11,16,17,20], which is widely used in artificial intelligence equipment, home security systems, environmental noise monitoring, etc. The continuous expansion of the dataset and further development of deep neural networks have made the ASC achieved better performance [2,24,29,31]. Most existing methods of ASC presume that the acoustic signal is collected from one of the predefined acoustic scenes called closed set, which may not be always true in a practical scenario. To overcome the limitation brought by presumption, Daniele et al. [3] introduce the open set setting into ASC. Compared with the closed set ASC that only needs to identify the known scenes, the

X. Shao et al. (Eds.): CSMT 2020, LNEE 761, pp. 93–104, 2021.
https://doi.org/10.1007/978-981-16-1649-5_8

open set ASC can not only classify audio clips to predefined acoustic scenes but also identify audio signals collected from the unknown acoustic scenes.

At present, few works focus on open set ASC [3,18,28,33]. Most methods [28, 33] usually regard the unknown acoustic scenes as a special known class and then obtain the prediction output by training of deep neural network. Finally, a "one-size-fits-all" fixed threshold as the basis is employed for identifying known scenes and unknown scenes. The threshold distinguishes the scene with a predicted probability higher than this value as a known class, and vice versa, as an unknown class. An obvious problem is that discriminative threshold is not involved in training phase. Therefore, the boundary between known scenes and unknown scenes is not learned actually.

To program the boundary between known scenes and unknown scenes, the open set acoustic scene detection is also regarded as a specialised adversarial domain adaptation problem as suggested by Saito et al. [21] and Fu et al. [7]. The open set adversarial domain adaptation methods have achieved excellent performance in the field of image recognition [9,23]. As an adversarial domain adaptation problem, training dataset and testing dataset can be considered as source domain and target domain respectively. This method uses the information learned in source domain to guide classification of known scenes in target domain. Moreover, the target domain distinguishes unknown class from known classes by adversarial training.

In this paper, we apply adversarial domain adaptation to the open set ASC, which solves the problem that the boundary between known scenes and unknown scenes cannot be truly learned. Because the "one-size-fits-all" threshold is not involved in the model training phase. Experiments show that our proposed method could effectively solve open set ASC.

2 Related Work

The purpose of closed set ASC is to match audio signals with predefined scene labels. Many methods benefit from the development of deep neural networks [1,4,6,19,26]. Hershey et al. [12] propose a Convolutional Neural Networks (CNNs) framework which performs well in image recognition for ASC. Vu et al. [26] employ Recurrent Neural Networks (RNNs) which is flexible in dealing with sequential data for ASC. Mun et al. [17] utilize Generative Adversarial Networks (GANs) to improve ASC performance by generating additional training dataset. Closed set ASC has achieved impressive performance to matching known semantic label with audio representing its recording environment. However, in practical applications, the processed dataset usually contains some samples, which are recorded in unknown scenarios outside. Recognizing a dataset incorporating unknown scene classes as above is called open set ASC. Considering the changes in dataset composition, traditional closed set ASC methods are no longer applicable to open set ASC. However, closed set ASC still provides reliable models for the classification of known scenes to open set ASC.

Open set ASC is closer to reality and more challenging compared with closed set ASC. Therefore, this paper will concentrate on the task of open set ASC, which is rarely studied. Battaglino et al. [3] first introduced the open set problem into the field of ASC. Zhu et al. [33] subdivide known classes through CNNs and self-attention mechanisms [25,30], and rely upon a fixed threshold to filter unknown scenes. Lei et al. [14] propose a method of combining an improved ResNet variant and a threshold for open set ASC. These methods are affected by closed set ASC on classifier, so they can classify known scenes well, and then identify the scene whose predicted probability is lower than the threshold as an unknown scene. It can be seen that, except for the final discrimination threshold, the entire classification network is highly similar to closed set ASC. The training process does not involve participation of this threshold, therefore, the classification network does not really program the boundary between known scenes and unknown scenes. And the "one-size-fits-all" threshold makes the system less robust and affected by the setting changes.

To avoid the problem of traditional methods, the open set ASC can be regarded as an adversarial domain adaptation problem which has achieved excellent performance in image recognition [15,22,27]. Adversarial domain adaptation consists of labelled source domain and unlabeled target domain which reduces the complexity of data labelling. The performance of target domain model is improved by transferring the knowledge learned from information-rich source domain model. To eliminate the difficulty of discriminating unknown class when aligning source domain and known classes in target domain, Saito et al. [21] propose to solve the open set domain adaptation by backpropagation in adversarial training. Based on the work of [21], Fu et al. [7] develop symmetrical Kullback Leibler (KL) distance to upgrade loss function of adversarial adaptation, so as to better recognize the potential unknown samples. These methods get rid of the dependence on "one-size-fits-all" threshold which not participating in training phase through means of adversarial domain adaptation.

Therefore, inspired by the adversarial domain adaptation in image recognition [21], we propose to introduce adversarial domain adaptation to solve the problem of open set ASC for the first time. The proposed method makes the model obtain the difference between known classes and unknown classes using pseudo-threshold by adversarial training, and remedies negative influence of the traditional threshold and improves the robustness of the system. Experiments demonstrate that the method proposed not only solves the problem that the threshold which not involved in training phase cannot learn the boundary between known classes and unknown classes in the traditional method but also can better classify the known scenes.

3 Open Set Adversarial Domain Adaptation

In this section, we first give an overview of the open set ASC system as shown in Fig. 1, and then detail the adversarial domain adaptation method and its training process.

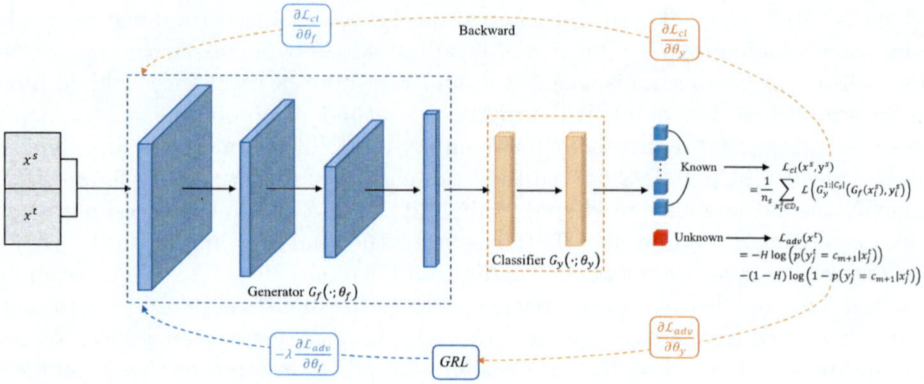

Fig. 1. Open set ASC system based on adversarial domain adaptation. The system is composed of a generator G_f and a classifier G_y. In target domain, the boundary between known classes and unknown classes is learned by GRL [8] inverting the gradient in an adversarial training way.

3.1 Problem Setting and System Framework

The open set adversarial domain adaptation is composed of source domain and target domain. Some known scenes are processed as source domain, and the other known scenes and unknown scenes are processed as target domain. In consideration of being close to the actual situation and relieving the pressure of labelling, target domain with unlabelled data is unsupervised. Here, source domain $\mathcal{D}_s$ and target domain $\mathcal{D}_t$ are denoted as $\{(x_i^s, y_i^s)\}_{i=1}^{n_s}$ and $\{x_j^t\}_{j=1}^{n_t}$ respectively, where n_s and n_t represent the number of samples in source domain and target domain. Compared with the set of classes $\mathcal{C}_s = \{c_1, c_2, ..., c_m\}_{m=1}^{K}$ of source domain, the set of scene classes $\mathcal{C}_t = \{c_1, c_2, ..., c_m; c_{m+1}\}_{m=1}^{K}$ in target domain adds an "unknown" class which related to undefined scenarios, where K refers to the number of defined scene classes. The data distributions of source domain and target domain are P and Q respectively, and they are unequal to each other.

Towards open set ASC, on the one hand, adversarial domain adaptation is to adapt known scenes of target domain to source domain for obtaining the knowledge of classification scenes; on the other hand, unknown scene can be identified in target domain through the discriminating mechanism to avoid the interference of the negative transfer.

With a view to achieve the above challenges, the adversarial domain adaptation model has a generator $G_f(x;\theta_f)$ and a classifier $y = G_y(G_f(x);\theta_y)$ in series as shown in Fig. 1. θ_f and θ_y respectively represent the parameters of corresponding mapping. For the source domain, the function of generator and classifier is to classify the known scenes as accurately as possible; for the target domain, the classifier makes the output probability of the unknown sample tighter the set demarcation H, while the generator which extracts advanced features is learned in the opposite direction to H, and the generator and classifier are adversarial

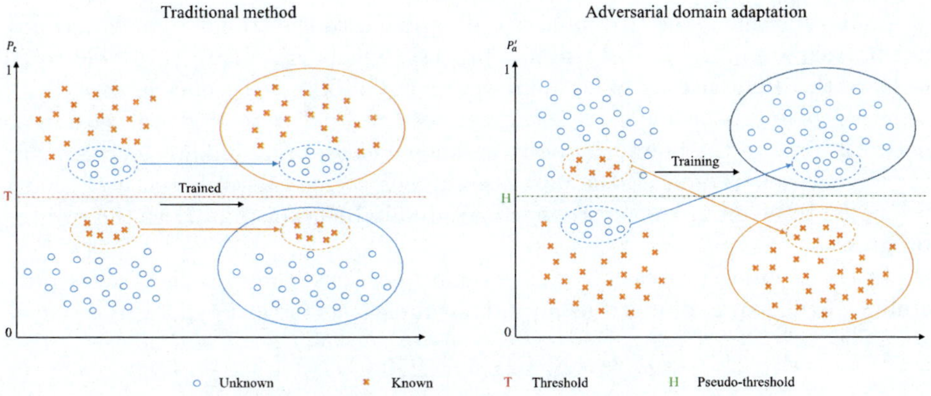

Fig. 2. The between the method proposed and the traditional method in identifying known and unknown classes. The traditional method uses a fixed threshold of T on the trained model while method proposed gradually distribute known and unknown classes to both sides of the preset pseudo-threshold H during the adversarial training process.

trained through a gradient reversal layer (GRL) [8] to complete the recognition task of unknown scenes.

3.2 Adversarial Domain Adaptation for Open Set ASC

Open set adversarial domain adaptation is to classify known scenes and identify unknown scenes in target domain at the same time. Considering that target domain is an unsupervised condition, source domain model would provide target domain with transferable knowledge of classification scenes. The generator G_f of source domain performs deep processing on the input x^s to extract advanced features related to the category difference. The classifier G_y completes the scene classification, obtains the probability prediction output p_s, and the loss function is defined as

$$\mathcal{L}_{cl} = \frac{1}{n_s} \sum_{x_i^s \in \mathcal{D}_s} \mathcal{L}(G_y^{1:K}(G_f(x_i^s)), y_i^s) \tag{1}$$

where $\mathcal{L}$ is the cross-entropy loss, G_y is a classifier for $K+1$ classes, $G_y^{1:K}$ represents the probability that the sample x_i^s is identified as known class with missing an unknown class than $\mathcal{C}_t$. Through the optimization of the loss of $\mathcal{L}_{cl}$, a network that can be transferred to target domain to classify known classes is finally obtained.

Then, we need to train a discrimination mechanism to regulate the boundary between known and unknown classes in target domain. Figure 2 shows the difference between the method proposed in this article and traditional method in identifying known and unknown classes. The traditional method usually uses a threshold to identify unknown class after the probability output P_t of the classifier. If P_t of known scene is lower than the threshold, this scene processed as an unknown scene. The threshold is only applied to the trained model and does not participate in the model training process, which causes this threshold to fail to learn the difference between unknown scenes and known scenes in actually. In order to solve this problem, we set a pseudo-threshold H to guide target domain network to learn the boundary between known classes and unknown classes during adversarial training. If the input is a known scene, the probability $P_a^{'}$ that is predicted to be an unknown scene will gradually be lower than H as the training progresses.

In the training process of target domain network, the classifier makes probability of the sample judged as an unknown class as far as possible to satisfy

$$p(y_j^t = c_{K+1}|x_j^t) = H, 0 < H < 1 \tag{2}$$

However, the generator is learned for keeping $p(y_j^t = c_{K+1}|x_j^t)$ as far away from H as possible. Such game training is completed by GRL [8] between the generator and the classifier. In the backpropagation process, the GRL enables the gradient of classification loss of the classifier automatically invert before backpropagating to the parameters of the generator as shown in Fig. 1, thereby realizing a adversarial training similar to GAN [10]. Among Fig. 1, the value of parameter λ is a hyperparameter. Correspondingly, the adversarial loss is defined as

$$\mathcal{L}_{adv} = -H\log(p(y_j^t = c_{K+1} \mid x_j^t)) - (1-H)\log(1 - p(y_j^t = c_{K+1} \mid x_j^t)) \tag{3}$$

With the adversarial training of generator and classifier, the probability that known scene in target domain is judged as an unknown class will be lower than H; conversely, the probability of the unknown sample being predicted as an unknown class will be higher than H. In this way, the samples of known classes and unknown classes will gradually treat the pseudo-threshold H as the boundary between each other during the training phase. Target domain can therefore achieve the purpose of recognizing unknown scenes.

3.3 Training Procedure

In the open set adversarial domain adaptation system, the network structure of source domain and target domain is same. Firstly, for guiding target domain to learn the knowledge of classifying known scenes, the generator and classifier obtain a network with excellent classification performance by training samples of the source domain. And the network is optimized by minimizing the loss $\mathcal{L}_{cl}$. Secondly, the target domain is able to identify unknown scenes, therefore, the generator and classifier learn the boundary between known scenes and unknown scenes in an adversarial manner through GRL. The loss that needs to be optimized is $\mathcal{L}_{adv}$. In short, the training goal of the entire model is

$$\begin{aligned} &\min_{G_y} \mathcal{L}_{cl} + \mathcal{L}_{adv} \\ &\min_{G_f} \mathcal{L}_{cl} - \mathcal{L}_{adv} \end{aligned} \tag{4}$$

Input	Generator				Classifier
Channels × Frames × Mel bins	CNN Block	CNN Block × 3	Flatten	FC × 2 LeakyReLU, BN	FC, Softmax
3 × 430 × 128	Kernel sizes: 5 × 5 Channel: 64 LeakyReLU, BN Pooling :2 × 2	Kernel sizes: 5 × 5 Channel: 64,128,128 LeakyReLU, BN Pooling :2 × 4			

Fig. 3. The proposed system network structure. "BN" is batch normalization, "FC" is fully connected layer.

Since source domain does not need to recognize unknown scenes, both the generator and classifier are trained in the direction of minimizing classification loss $\mathcal{L}_{cl}$; while target domain is different, generator and classifier are trained in the opposite direction of optimization, and then the pseudo-threshold H is turned into a true boundary through an adversarial method. Through the above training process, the target domain can not only classify known scenes but also identify unknown scenes without supervision.

4 Experiments and Results

4.1 Datasets and Experimental Setup

We evaluate the proposed method on the development dataset released by Detection and Classification on Acoustic Scenes and Events challenge 2019 (DCASE 2019) Task1 Subtask3. This dataset is a collection of audio scenes recorded in 12 European cities by professional recording equipment. The dataset consists of 10 known scenes and 1 unknown scene, where unknown scene includes 4 sub-scenes. The unknown class in target domain is related to three sub-scenes, and the unknown class in testing set is remaining sub-scene. Each audio sample has a duration of 10 s and is sampled to 44.1 kHz. As the input of our system, log mel spectrograms complemented by its deltas and delta-deltas were used. The input features are of size 430 time samples and 128 mel filter banks. In order to further expand data and improve system performance, we use Mixup [32] as a device to augment data.

This paper proposes an adversarial domain adaptation method for open set ASC. This method not only classifies known scenes like closed set ASC but also distinguishes unknown scenes from known scenes. As shown in Fig. 3, the adversarial domain adaptation system proposed consists of a generator and a classifier. Among them, the generator is used to extract advanced feature representations of input audio features, including 4 convolutional layer with kernel sizes of 5×5 and 2 fully connected layers. In order to stabilize training, we apply LeakyReLU and batch normalization after each convolutional layer. And average pooling is used for downsampling after convolutional layer. The classifier is composed of a fully connected layer, and the probability prediction output is obtained through Softmax. Stochastic gradient descent (SGD) [5] optimizer whose learning rate is 0.001 is used in this system.

Scene Class	Traditional method		OS-AND
	DCASE2019 Baseline	CNN-threshold [33]	
Airport	44.2	48.1	**62.2**
Shopping mall	50.9	52	**60.2**
Metro station	41.3	**59.3**	57.3
Street pedestrian	47.5	35.4	**69**
Public square	34.7	39.1	**51.1**
Street traffic	78.4	78.1	**85.2**
Tram	60.7	**70.2**	60
Bus	59.3	60.6	**72.1**
Metro	51.5	56.5	**69**
Park	74	**81.7**	81.2
Known Average	54.3	58.1	**66.7**
Unknown	43.1	48.1	**60.6**

Fig. 4. The comparison of accuracy (%) of the scene class obtained by mentioned model and traditional methods.

4.2 Performance and Analysis

The following Fig. 4 shows the comparison of experimental results between open set adversarial domain adaptation method (OS-AND) proposed in this paper and traditional methods. Baseline is a method officially released by DCASE 2019 Task1 Subtask3. The system consists of 2 convolutional layers and one dense layer. Finally, a threshold of 0.5 is used as the boundary for identifying unknown scenes. The method "CNN-threshold" proposed by Kong [13] uses a 5-layer CNN with a convolution kernel of 5×5 and a threshold of 0.5. It can be seen from the table that our model is significantly higher than traditional methods in the prediction results of known scenes and unknown scenes. In particular, the accuracy of unknown class has increased from 48.1% to 60.6%. The great improvement of the accuracy of unknown class recognition shows that the method of adversarial domain adaptation proposed in this paper is effective and reasonable. More than half of known scenes have achieved better accuracy than traditional methods and the average accuracy of known scenes is improved by about 9% compared with "CNN-threshold". This proves that the open set ASC based on adversarial domain adaptation proposed in this paper can not only identify unknown scenes but also classify known scenes well. The method proposed is more sensitive to scenes such as "Airport", "Street pedestrian", and "Metro" when classifying known scenes according to the experimental results. Of course, there are some scenes whose accuracy is lower than traditional methods, such as "Tram", "Park", etc. This is understandable. The target domain under unsupervised condition has some difficult samples, which will be processed as unknown scenes because of the low probability of similarity with known scenes in the process of adversarial training.

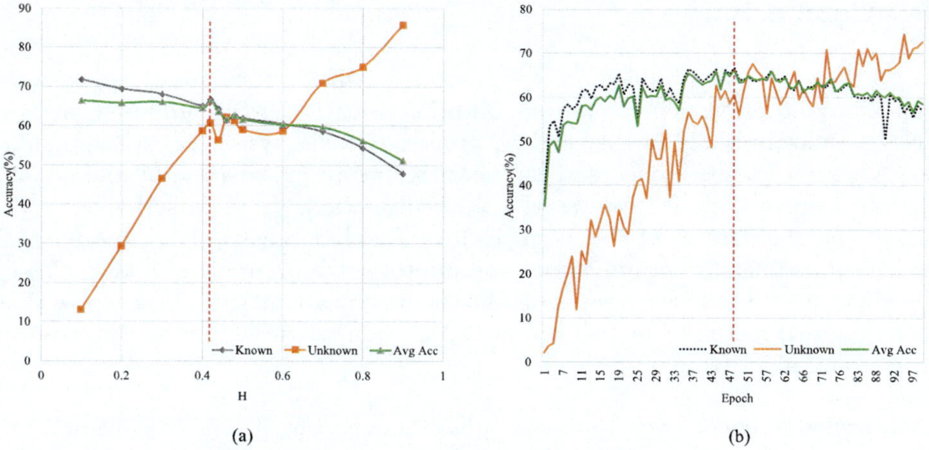

Fig. 5. (a) shows the variation curve of model accuracy with H; (b) shows the variation curve of model accuracy with epoch.

In addition, in order to better explain the performance of adversarial domain adaptation, we also give the relationship curve of the pseudo-threshold H of adversarial loss $\mathcal{L}_{adv}$ and accuracy, and the variation of the system accuracy with the training epoch under the optimal H in Fig. 5. The function of pseudo-threshold H in $\mathcal{L}_{adv}$ is to provide a quasi-differential boundary between the known and unknown classes of the target domain. The choice of H should ensure that the unknown class can be well recognized, and the classification accuracy of the known classes is not excessively lost. As shown in Fig. 5(a), the accuracy of the unknown class gradually increases as H increases, and the accuracy of known classes gradually decreases. However, the accuracy of known scenes has a small drop since the source domain provides the target domain with transferable scene classification knowledge. When $H = 0.42$ (as shown by the red dotted line in Fig. 5(a)), the average accuracy of the system is the highest. At this time, the accuracy of the unknown class increases greatly without sacrificing the accuracy of known classes too much. Therefore, H is chosen to be 0.42 in adversarial loss of $\mathcal{L}_{adv}$.

When $H = 0.42$, we plot the change of system accuracy to training epoch in Fig. 5(b). It can be seen that in the first 50 epochs of training, the accuracy of both unknown class and known classes fluctuates and rises, and the optimal result is reached at the 48th epoch (as shown by the red dotted line in Fig. 5(b)). After more than 50 epochs, although adversarial training increases the accuracy of unknown class, it also causes the accuracy of known classes to drop significantly. In summary, in order to enable model to identify unknown scenes and classify known scenes to achieve better performance, we choose the model obtained when $H = 0.42$ and the 48th training epoch.

5 Conclusion

In this paper, we propose an open set ASC method based on adversarial domain adaptation. Different from traditional methods, the proposed method uses pseudo-threshold in the training process to make system learn the boundary between known scenes and unknown scenes in an adversarial manner. In order to prove the effectiveness of proposed method, we set a series of experiments on the dataset of the DCASE 2019 Task1 Subtask3. The experimental results show that the proposed method improves the accuracy of known classes by about 9% and the accuracy of unknown class by about 18%. This proves that our proposed method can well complete the open set ASC of classifying known scenes and identifying unknown scenes.

Acknowledgements. This work was supported in part by the National Natural Science Foundation of China (62001038).

References

1. Bae, S.H., Choi, I., Kim, N.S.: Acoustic scene classification using parallel combination of LSTM and CNN. In: Proceedings of the Detection and Classification of Acoustic Scenes and Events 2016 Workshop (DCASE2016), pp. 11–15 (2016)
2. Barchiesi, D., Giannoulis, D., Stowell, D., Plumbley, M.D.: Acoustic scene classification: classifying environments from the sounds they produce. IEEE Signal Process. Mag. **32**(3), 16–34 (2015)
3. Battaglino, D., Lepauloux, L., Evans, N.: The open-set problem in acoustic scene classification. In: 2016 IEEE International Workshop on Acoustic Signal Enhancement (IWAENC), pp. 1–5. IEEE (2016)
4. Bisot, V., Serizel, R., Essid, S., Richard, G.: Leveraging deep neural networks with nonnegative representations for improved environmental sound classification. In: IEEE International Workshop on Machine Learning for Signal Processing (2017)
5. Bottou, L.: Large-scale machine learning with stochastic gradient descent. In: Proceedings of COMPSTAT 2010, pp. 177–186. Springer (2010)
6. Eghbal-zadeh, H., Lehner, B., Dorfer, M., Widmer, G.: A hybrid approach with multi-channel i-vectors and convolutional neural networks for acoustic scene classification. In: 2017 25th European Signal Processing Conference (EUSIPCO), pp. 2749–2753 (2017)
7. Fu, J., Wu, X., Zhang, S., Yan, J.: Improved open set domain adaptation with backpropagation. In: 2019 IEEE International Conference on Image Processing (ICIP), pp. 2506–2510. IEEE (2019)
8. Ganin, Y., Lempitsky, V.: Unsupervised domain adaptation by backpropagation. In: International Conference on Machine Learning, pp. 1180–1189 (2015)

9. Ganin, Y., Ustinova, E., Ajakan, H., Germain, P., Larochelle, H., Laviolette, F., Marchand, M., Lempitsky, V.: Domain-adversarial training of neural networks. In: Domain Adaptation in Computer Vision Applications, pp. 189–209. Springer (2017)
10. Goodfellow, I., Pouget-Abadie, J., Mirza, M., Xu, B., Warde-Farley, D., Ozair, S., Courville, A., Bengio, Y.: Generative adversarial nets. In: Advances in Neural Information Processing Systems, pp. 2672–2680 (2014)
11. Heittola, T., Mesaros, A.: DCASE 2017 challenge setup: tasks datasets and baseline system. Technical report, DCASE 2017 Challenge (2017)
12. Hershey, S., Chaudhuri, S., Ellis, D.P.W., Gemmeke, J.F., Jansen, A., Moore, R.C., Plakal, M., Platt, D., Saurous, R.A., Seybold, B., Slaney, M., Weiss, R.J., Wilson, K.: CNN architectures for large-scale audio classification. In: 2017 IEEE International Conference on Acoustics, Speech and Signal Processing (ICASSP), pp. 131–135 (2017)
13. Kong, Q., Cao, Y., Iqbal, T., Xu, Y., Wang, W., Plumbley, M.D.: Cross-task learning for audio tagging, sound event detection and spatial localization: DCASE 2019 baseline systems. arXiv preprint arXiv:1904.03476 (2019)
14. Lei, C., Wang, Z.: Multi-scale recalibrated features fusion for acoustic scene classification technical report. DCASE 2019 Challenge, Technical report (2019)
15. Long, M., Cao, Y., Wang, J., Jordan, M.I.: Learning transferable features with deep adaptation networks. In: Proceedings of the 32nd International Conference on International Conference on Machine Learning, ICML 2015, vol. 37. pp. 97–105. JMLR.org (2015)
16. Mesaros, A., Heittola, T., Benetos, E., Foster, P., Lagrange, M., Virtanen, T., Plumbley, M.D.: Detection and classification of acoustic scenes and events: outcome of the DCASE 2016 challenge. IEEE/ACM Trans. Audio Speech Lang. Proces. (TASLP) **26**(2), 379–393 (2018)
17. Mun, S., Park, S., Han, D.K., Ko, H.: Generative adversarial network based acoustic scene training set augmentation and selection using SVM hyper-plane. In: Proceedings of the DCASE, pp. 93–97 (2017)
18. Rakowski, A., Kosmider, M.: Frequency-aware CNN for open set acoustic scene classification. DCASE 2019 Challenge, Technical report (2019)
19. Ren, Z., Kong, Q., Han, J., Plumbley, M.D., Schuller, B.W.: Attention-based atrous convolutional neural networks: visualisation and understanding perspectives of acoustic scenes. In: 2019 IEEE International Conference on Acoustics, Speech and Signal Processing (ICASSP), ICASSP 2019, pp. 56–60. IEEE (2019)
20. Ren, Z., Qian, K., Wang, Y., Zhang, Z., Pandit, V., Baird, A., Schuller, B.: Deep scalogram representations for acoustic scene classification. IEEE/CAA J. Autom. Sin. **5**(3), 662–669 (2018)
21. Saito, K., Yamamoto, S., Ushiku, Y., Harada, T.: Open set domain adaptation by backpropagation. In: Proceedings of the European Conference on Computer Vision (ECCV), pp. 153–168 (2018)
22. Scheirer, W.J., de Rezende Rocha, A., Sapkota, A., Boult, T.E.: Toward open set recognition. IEEE Trans. Pattern Anal. Mach. Intell. **35**(7), 1757–1772 (2012)
23. Sun, B., Saenko, K.: Deep coral: correlation alignment for deep domain adaptation. In: Hua, G., Jégou, H. (eds.) Computer Vision - ECCV 2016 Workshops, pp. 443–450. Springer International Publishing, Cham (2016)
24. Valenti, M., Diment, A., Parascandolo, G., Squartini, S., Virtanen, T.: DCASE 2016 acoustic scene classification using convolutional neural networks. In: Proceedings of the Workshop Detection Classification of Acoustic Scenes and Events, pp. 95–99 (2016)

25. Vaswani, A., Shazeer, N., Parmar, N., Uszkoreit, J., Jones, L., Gomez, A.N., Kaiser, Ł., Polosukhin, I.: Attention is all you need. In: Advances in Neural Information Processing Systems, pp. 5998–6008 (2017)
26. Vu, T.H., Wang, J.C.: Acoustic scene and event recognition using recurrent neural networks. In: Detection and Classification of Acoustic Scenes and Events 2016 (2016)
27. Wan, C.H., Chuang, S.P., Lee, H.Y.: Towards audio to scene image synthesis using generative adversarial network. In: 2019 IEEE International Conference on Acoustics, Speech and Signal Processing (ICASSP), ICASSP 2019, pp. 496–500. IEEE (2019)
28. Wilkinghoff, K., Kurth, F.: Open-set acoustic scene classification with deep convolutional autoencoders (2019)
29. Wu, Y., Lee, T.: Enhancing sound texture in CNN-based acoustic scene classification. In: ICASSP 2019-2019 IEEE International Conference on Acoustics, Speech and Signal Processing (ICASSP), pp. 815–819. IEEE (2019)
30. Xu, Y., Kong, Q., Wang, W., Plumbley, M.D.: Large-scale weakly supervised audio classification using gated convolutional neural network. In: 2018 IEEE International Conference on Acoustics, Speech and Signal Processing (ICASSP), pp. 121–125. IEEE (2018)
31. Yang, Y., Zhang, H., Tu, W., Ai, H., Cai, L., Hu, R., Xiang, F.: Kullback–Leibler divergence frequency warping scale for acoustic scene classification using convolutional neural network. In: 2019 IEEE International Conference on Acoustics, Speech and Signal Processing (ICASSP), ICASSP 2019, pp. 840–844. IEEE (2019)
32. Zhang, H., Cisse, M., Dauphin, Y.N., Lopez-Paz, D.: mixup: beyond empirical risk minimization. arXiv preprint arXiv:1710.09412 (2017)
33. Zhu, H., Ren, C., Wang, J., Li, S., Wang, L., Yang, L.: DCASE 2019 challenge task1 technical report. DCASE 2019 Challenge, Technical report (2019)

Check for updates

Active Room Compensation for 2.5D Sound Field Reproduction

Yitong Chen(✉) and Wen Zhang

Center of Intelligent Acoustics and Immersive Communications, Northwestern Polytechnical University, Xi'an 710072, China
wen.zhang@nwpu.edu.cn
https://www.researchgate.net/profile/Wen_Zhang20

Abstract. Modelling secondary sources as 3D point sources to reproduce 2D desired sound field is named as 2.5D sound field reproduction, which has the intrinsic dimensionality mismatch problem. Existing methods for 2.5D reproduction have focused on solving the dimensionality mismatch problem and mostly considered free-field condition. However, in most cases, the reverberation caused by the listening room will degrade the reproduction performance. In this work, we propose an active room compensation strategy for 2.5D reproduction. Firstly, adopt sectorial mode matching algorithm to achieve 2.5D reproduction, the desired sound field and generated sound field are matched at the reproduction center. Secondly, the modal-domain algorithm is developed to estimate reverberant sound field and design the compensation signals of loudspeakers to compensate reverberant sound field. The proposed method is validated through simulation experiments to demonstrate its effectiveness against room reverberations.

Keywords: Sound field reproduction · 2.5D reproduction · Room compensation · Modal domain

1 Introduction

As psychoacoustics theory puts the fact that human ears are more sensitive to sounds from horizontal direction at the height of ears, it is more valuable to reproduce desired sound field on two-dimension field. Early studies regard it as 2D reproduction problem where secondary sources are modelled as vertical line sources [1]. However, due to the acoustic characteristics of the loudspeaker, modelling it as 3D point source is more reasonable in practice. Due to the intrinsic dimensionality mismatch, modelling secondary sources as 3D point sources to reproduce 2D sound field is named as 2.5D reproduction [2–5]. Most existing methods for 2.5D reproduction can make good performance in addressing dimensionality mismatch problem [2,6]. However, these approaches focus on free-field condition while the reverberation caused by time-varying room responses will impair the generated sound field [1]. Thus, room compensation (or room equalization) has been proposed to reduce or eliminate the reverberant effects.

X. Shao et al. (Eds.): CSMT 2020, LNEE 761, pp. 105–112, 2021.
https://doi.org/10.1007/978-981-16-1649-5_9

As for room compensation, current proposed methods can be divided into two ways: passive compensation manners and active manners. Passive compensation methods reduce wall reflections by acoustic absorption materials. Nevertheless, it gets costly and impractical especially at low frequencies in many real-world application scenarios. Thus, researchers put forward compensation methods by active manners. In order to equalize the effects of reverberation, adding appropriate compensation signals on the loudspeaker arrays is the core. Currently, most proposed room compensation techniques are based on Multiple Input Multiple Output (MIMO) system [7,8], which can only reduce the effect of reverberation at discrete points and its adjacent region. Designing compensation signals using modal domain processing [9] achieves compensation within a continuous region. However, this has only been considered for 2D reproduction.

In this paper, we propose an active room compensation approach in 2.5D reproduction through modal domain processing. Section 2 reviews the 2.5D reproduction using sectorial mode matching algorithm, and we propose an active room compensation algorithm in Sect. 3. In Sect. 4, the proposed algorithm is simulated and evaluated by reproduction of narrowband and broadband signals under reverberant environment.

2 2.5D Reproduction

2.1 Problem Formulation

In modal-domain, we express the 2D desired sound field at any point $\boldsymbol{x} = \{r_x, \phi_x\}$ through the interior solution of wave equation [10]

$$P_d(\boldsymbol{x}, k) \approx \sum_{m=-M}^{M} \alpha_m(k) J_m(kr_x) e^{im\phi_x}, \tag{1}$$

where k represents the wave number, $\alpha_m(k)$ represents sound field coefficients, $J_m(\cdot)$ represents cylindrical Bessel function. The truncation order is determined as $M = \lceil ekR/2 \rceil$ [11] where R is the radius of control region.

Consider adopting circular loudspeaker array to generate the desired sound field exactly as (1) shows, where the array is implemented on horizontal plane, the number of loudspeakers has to satisfy $L \geq 2M + 1$ [12].

Thus, reproduced sound field in free-field is formulated as

$$P(\boldsymbol{x}, k) = \sum_{l=1}^{L} d_l(k) H_l(\boldsymbol{x}, k), \tag{2}$$

where $H_l(\boldsymbol{x}, k)$ is acoustic transfer function (ATF) of the lth loudspeaker and the observation spot x, $d_l(k)$ is the driving signal of lth loudspeaker.

The expression of $H_l(\boldsymbol{x}, k)$ is as follows

$$H_l(\boldsymbol{x}, k) = \frac{e^{-ik\|\boldsymbol{y}_l - \boldsymbol{x}\|}}{4\pi\|\boldsymbol{y}_l - \boldsymbol{x}\|}, \tag{3}$$

where $\boldsymbol{y}_l = \{r_l, \phi_l\}$ represents the loudspeaker location and $\|\cdot\|$ denotes the Euclidean distance of the vectors.

In modal domain, $H_l(\boldsymbol{x}, k)$ is formulated as

$$H_l(\boldsymbol{x}, k) \approx \sum_{m=-M}^{M} \sum_{n=|m|}^{M} \gamma_n^m(l,k) j_n(kr_x) Y_n^m(\frac{\pi}{2}, \phi_x), \tag{4}$$

where $\gamma_n^m(l,k)$ is the ATF coefficient of lth source. In free field condition,

$$\gamma_n^m(l,k) = -ikh_n^{(2)}(kr_l)\overline{Y_n^m(\frac{\pi}{2}, \phi_l)},$$

Substituting (4) into (2) gives the expression

$$P(\boldsymbol{x}, k) \approx \sum_{m=-M}^{M} \sum_{l=1}^{L} d_l(k) \sum_{n=|m|}^{M} \gamma_n^m(l,k) j_n(kr_x) Y_n^m e^{im\phi_x}, \tag{5}$$

where the spherical harmonic function at elevation $\theta = \pi/2$ is defined as $Y_n^m(\frac{\pi}{2}, \phi_x) = Y_n^m e^{im\phi_x}$, with $Y_n^m \triangleq A_n^m P_n^m(0)$, $A_n^m = \sqrt{\frac{2n+1}{4\pi} \frac{(n-|m|)!}{(n+|m|)!}}$ and $P_n^m(\cdot)$ represents associated Legendre function.

The loudspeaker driving signals are designed by equating (1) with (5). Using the orthogonality property of complex exponentials, it gives the following equation

$$\alpha_m(k) J_m(kr_x) = \sum_{l=1}^{L} d_l(k) \underbrace{\sum_{n=|m|}^{M} \gamma_n^m(l,k) j_n(kr_x) Y_n^m}_{h_m(l,k,r_x)}, \tag{6}$$

for $m = -M, ..., M$. From (6), it shows that the expansion is only over mode m and the Bessel function $J_m(kr_x)$ denotes the radial propagation in 2D sound field, and the modal expansion is over both n and m and the spherical Bessel function $j_n(kr_x)$ denotes the radial propagation in 3D sound field, which reflects the dimensionality mismatching. The solution of this problem is finding appropriate matching distance r_x.

2.2 Sectorial Mode Matching

It has been proved [2] that adopting the center of reproduction region as the matching spot ($r_x = 0$) can reach a relatively less distortion due to dimensionality mismatch. Note that when $kr_x \to 0$, the summation over order n in (4) reduces to the single term $n = |m|$ [2,13] where $Y_{|m|}^m(\frac{\pi}{2}, \phi_x)$ is termed as the sectorial harmonics, and $h_m(l,k,r_x)$ in (6) is expressed as

$$h_m(l,k,r_x) = \gamma_{|m|}^m(l,k) j_{|m|}(kr_x) Y_{|m|}^m, \tag{7}$$

Thus, the driving signals $d_l(k)$ are derived by matrix

$$\mathbf{d} = \mathbf{H}^{\dagger}\mathbf{b}, \tag{8}$$

where

$$\mathbf{d} = \left[d_{l_1}(k) \cdots d_{l_L}(k)\right]^T,$$

$$\mathbf{H}^{\dagger} = \begin{bmatrix} h_{-M}(l_1, k, r_x) & \ldots & h_{-M}(l_L, k, r_x) \\ \vdots & \ddots & \vdots \\ h_M(l_1, k, r_x) & \ldots & h_M(l_L, k, r_x) \end{bmatrix}^{\dagger},$$

$$\mathbf{b} = \left[\alpha_{-M}(k)J_{-M}(kr_x) \cdots \alpha_M(k)J_M(kr_x)\right]^T.$$

The dependence on wavenumber k is omitted for notation simplicity. Then, driving signals can be expressed simply as

$$\mathbf{d} = (\mathbf{H}^{\mathbf{d}})^{\dagger}\mathbf{b}, \tag{9}$$

where $\mathbf{H}^{\mathbf{d}}$ denotes the direct-path loudspeaker array ATF in modal domain.

3 Active Room Compensation

The previous section reviews sectorial mode matching algorithm for 2.5D reproduction. This section introduces the active room compensation algorithm for 2.5D reproduction in modal domain.

We use a circular microphone array of Q microphones to encircle the control region with the microphones uniformly placed. Figure 1 shows the system setup.

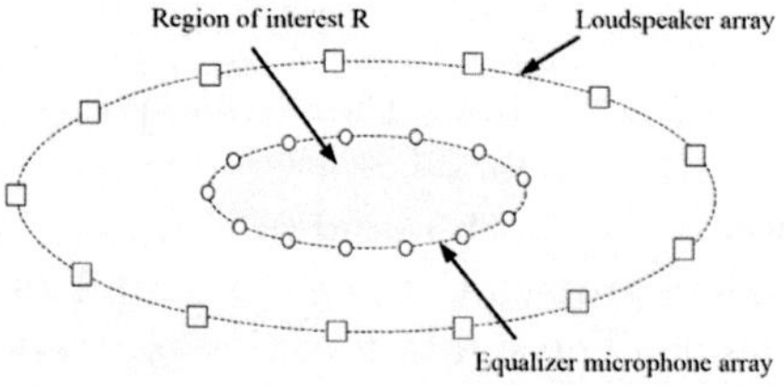

Fig. 1. Active room compensation system setup: the loudspeaker array and a plane control region circled by equalizer microphone array.

Note that the aim here is to compensate the reverberation caused by the loudspeaker signals. Based on Eq. (7) and given the fact that only the sectorial modes are controlled, derive the sound field coefficients measured by equalizer microphone array

$$\beta_m(k) = \frac{1}{Qj_{|m|}(kr_M)Y_{|m|}^m}\sum_{q=1}^{Q} P(\boldsymbol{x}_q, k)e^{-im\phi_q}, \qquad m = -M, \ldots, M, \tag{10}$$

where r_M denotes the microphone array radius. In vector form, the measured sound field coefficients are constituted by direct-path modes and reverberant-path modes

$$\boldsymbol{\beta} = \boldsymbol{\beta}^{\mathbf{d}} + \boldsymbol{\beta}^{\mathbf{r}}, \tag{11}$$

The modal domain sound field coefficients can also be expressed as the loudspeaker driving signals and the corresponding ATF coefficients

$$\boldsymbol{\beta} = \boldsymbol{\Gamma}\mathbf{d} = \boldsymbol{\Gamma}^{\mathbf{d}}\mathbf{d} + \boldsymbol{\Gamma}^{\mathbf{r}}\mathbf{d}, \tag{12}$$

where $\boldsymbol{\Gamma} = [\boldsymbol{\gamma}_1, \ldots, \boldsymbol{\gamma}_L]$, and $\boldsymbol{\gamma}_l = [\gamma_{|M|}^{-M}(l,k), \ldots, \gamma_{|M|}^{M}(l,k)]^T$ which represent the ATF coefficients matrix. The ATF coefficients matrix is also decomposed into the direct path and reverberant path, that is $\boldsymbol{\Gamma} = \boldsymbol{\Gamma}^{\mathbf{d}} + \boldsymbol{\Gamma}^{\mathbf{r}}$.

Then, the compensation problem is formulated as introducing the compensation signals $\boldsymbol{\delta}\mathbf{d}$ at the loudspeakers to minimize the reverberant-path modes, i.e.,

$$\begin{aligned} \min \|\boldsymbol{\beta}^{\mathbf{r}}\| &= \|\boldsymbol{\beta} - \boldsymbol{\beta}^{\mathbf{d}}\| \\ &= \|\boldsymbol{\Gamma}(\mathbf{d} + \boldsymbol{\delta}\mathbf{d}) - \boldsymbol{\Gamma}^{\mathbf{d}}\mathbf{d}\| \\ &= \|(\boldsymbol{\Gamma}^{\mathbf{d}} + \boldsymbol{\Gamma}^{\mathbf{r}})\boldsymbol{\delta}\mathbf{d} + \boldsymbol{\Gamma}^{\mathbf{r}}\mathbf{d}\| \end{aligned} \tag{13}$$

Thus, by solving (13) in the least-squares approach, the compensation signals at the loudspeakers for eliminating reverberation is given by

$$\boldsymbol{\delta}\mathbf{d} = -(\boldsymbol{\Gamma}^{\mathbf{d}} + \boldsymbol{\Gamma}^{\mathbf{r}})^{\dagger}(\boldsymbol{\beta} - \boldsymbol{\Gamma}^{\mathbf{d}}\mathbf{d}), \tag{14}$$

where the measured sound field coefficients $\boldsymbol{\beta}$, the loudspeaker driving signals $\mathbf{d}$ for direct path (or free-field) propagation $\boldsymbol{\Gamma}^{\mathbf{d}}$ can be accessed straightforwardly. The reverberant path ATF coefficients are obtained in an adaptive manner as addressed in [9].

4 Evaluation

This section describes the simulation based experimental setup and performance of the proposed room compensation algorithm.

4.1 Simulation Results

Consider a 2D reverberant room of size $2\,\mathrm{m} \times 2\,\mathrm{m}$. Simulate reverberant environment by image source method [14]: the wall reflection is 0.8 (the ceiling and floor are perfectly-absorbing) and the image depth of 7. Control region circled by equalizer microphone array is centred at (1 m, 1 m) with the radius of 0.25 m. The loudspeaker array is uniformly placed on the circle of 1 m radius. Note that the number of loudspeakers and microphones are both 39, which satisfies the mode truncation requirement $M = \lceil ekR/2 \rceil + 1$ [13].

We simulate reproduction of two kinds of narrowband sources, i.e., plane wave and cylindrical wave, and a broadband source. For the narrowband cases, the plane wave source is of 3 kHz frequency, incident from $\phi_v = \pi/3$ and the cylindrical wave source is of 3 kHz frequency, locating at $r_v = 1.5$ m, $\phi_v = \pi/3$ away from the system center. For the broadband cases, the cylindrical wave source is of 3 kHz bandwidth: frequencies 100 Hz to 3 kHz, and the location is as same as narrowband cylindrical wave source. Figure 2 and Fig. 3 show the compensation performance of narrowband sources, where the desired, reverberant and compensated sound fields are displayed in (a), (b), (c) respectively. Compared with the reverberant sound fields in Fig. 2(b) and Fig. 3(b), compensation is achieved within the whole reproduction region.

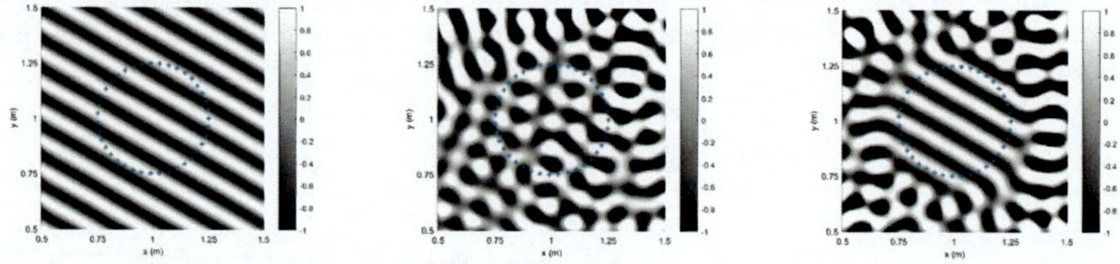

(a) (b) (c)

Fig. 2. Reproduction of a plane wave (frequency at 3 kHz, incident from $\phi_v = \pi/3$) in a 0.25 m control region circled by the asterisk

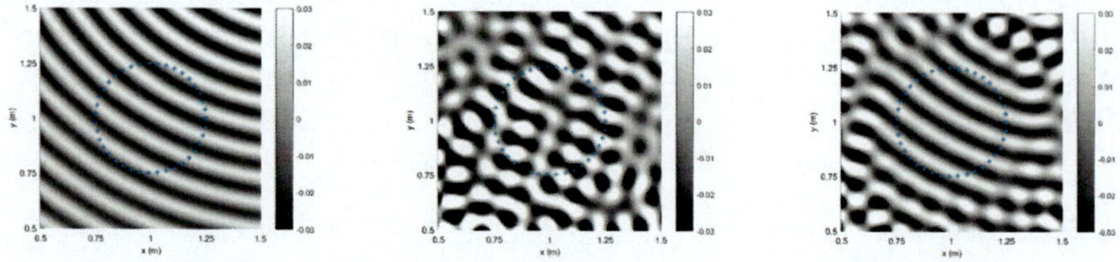

(a) (b) (c)

Fig. 3. Reproduction of a cylindrical wave (frequency at 3 kHz, locating at $r_v = 1.5$ m, $\phi_v = \pi/3$ away from the system center) in a 0.25 m control region circled by the asterisk

4.2 Evaluation of Results

The compensation performance is measured by the normalized equalization error ε over the whole control region. The normalized equalization error is defined as follows,

$$\varepsilon(w) = 10 \log_{10} \frac{\int_R |P(\boldsymbol{x}, k) - P_d(\boldsymbol{x}, k)|^2 d\boldsymbol{x}}{\int_R |P_d(\boldsymbol{x}, k)|^2 d\boldsymbol{x}} \tag{15}$$

In this simulation, 40000 observation points are uniformly selected within the reproduction region to approximate the integral.

In broadband case, set the number of loudspeakers as 45, which is slightly more than the number of loudspeakers designed at the maximum frequency 3 kHz, other simulation settings are the same as in the examples of Fig. 2 and Fig. 3. In Fig. 4, it demonstrates the normalized equalization error over 3 kHz bandwidth, which shows that setting the number of the loudspeakers more than the minimum requirement of certain frequency can achieve satisfying results. As the frequency increases, the performance will gradually degrade.

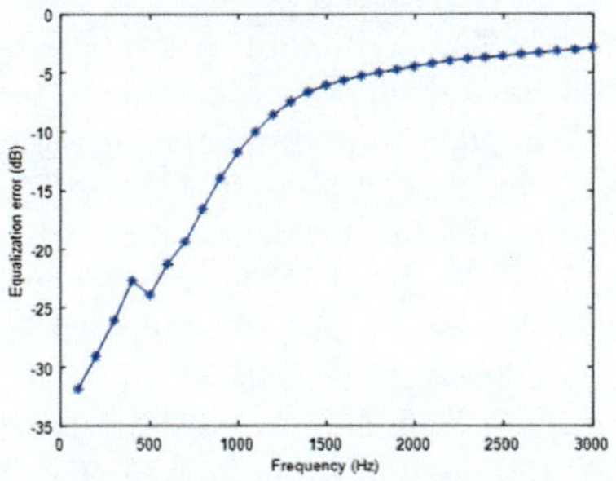

Fig. 4. The normalized equalization error of cylindrical wave reproduction over 3 kHz bandwidth frequency range.

5 Conclusion

In this paper, we propose an active compensation algorithm for 2.5D sound field reproduction. Firstly, the sectorial mode matching method is used to derive the loudspeaker driving signals assuming direct-path propagation between the loudspeaker array and reproduction region. Then, based on sectorial mode, an active compensation algorithm is introduced to compensate room reverberation. From simulation results, it can proved that the proposed algorithm can achieve effective room compensation over the whole reproduction region.

References

1. Betlehem, T., Abhayapala, T.D.: Theory and design of sound field reproduction in reverberant rooms. J. Acoust. Soc. Am. **117**(4), 2100–2111 (2005). https://doi.org/10.1121/1.1863032
2. Zhang, W., Abhayapala, T.D.: 2.5D sound field reproduction in higher order Ambisonics. In: 14th IEEE International Workshop on Acoustic Signal Enhancement, pp. 342–346. IEEE Press, Juan-les-Pins (2014). https://doi.org/10.1109/iwaenc.2014.6954315

3. Zhang, W., Zhang, J., Abhayapala, T.D., Zhang, L.: 2.5D multizone reproduction using weighted mode matching. In: 2018 IEEE International Conference on Acoustics, Speech and Signal Processing, pp. 476–480. IEEE Press, Calgary (2018). https://doi.org/10.1109/ICASSP.2018.8462511
4. Okamoto, T.: 2.5D higher order ambisonics for a sound field described by angular spectrum coefficients. In: 2016 IEEE International Conference on Acoustics, Speech and Signal Processing, pp. 326–330. IEEE Press, Shanghai (2016). https://doi.org/10.1109/ICASSP.2016.7471690
5. Okamoto, T.: Horizontal 3D sound field recording and 2.5D synthesis with omnidirectional circular arrays. In: 2019 IEEE International Conference on Acoustics, Speech and Signal Processing, pp. 960–964. IEEE Press, Brighton (2019). https://doi.org/10.1109/ICASSP.2019.8683009
6. Wang, W., Jia, M., Bao, C., Zhang, J.: 2.5D interior/exterior sound field reproduction and its extension to narrowband speech signals. In: 2016 IEEE International Conference on Audio, Language and Image Processing, pp. 7–12. IEEE Press, Shanghai (2016). https://doi.org/10.1109/ICALIP.2016.7846548
7. Brännmark, L.: Robust audio precompensation with probabilistic modeling of transfer function variability. In: 2009 IEEE Workshop on Applications of Signal Processing to Audio and Acoustics, pp. 193–196. IEEE Press, New Paltz (2009). https://doi.org/10.1109/ASPAA.2009.5346533
8. Brännmark, L., Bahne, A., Ahlén, A.: Compensation of loudspeaker-room responses in a robust MIMO control framework. IEEE-ACM Trans. Audio Speech Lang. **21**(6), 1201–1216 (2013). https://doi.org/10.1109/TASL.2013.2245650
9. Talagala, D.S., Zhang, W., Abhayapala, T.D.: Efficient multi-channel adaptive room compensation for spatial soundfield reproduction using a modal decomposition. IEEE-ACM Trans. Audio Speech Lang. **22**(10), 1522–1532 (2014). https://doi.org/10.1109/TASLP.2014.2339195
10. Williams, E.G.: Fourier Acoustics: Sound Radiation and Nearfield Acoustical Holography. Academic, New York (1999)
11. Kennedy, R.A., Sadeghi, P., Abhayapala, T.D., Jones, H.M.: Intrinsic limits of dimensionality and richness in random multipath fields. IEEE Trans. Signal Process. **55**(6), 2542–2556 (2007). https://doi.org/10.1109/TSP.2007.893738
12. Ward, D.B., Abhayapala, T.D.: Reproduction of a plane-wave sound field using an array of loudspeakers. IEEE-ACM Trans. Audio Speech Lang. **9**(6), 697–707 (2001). https://doi.org/10.1109/89.943347
13. Poletti, M.A., Betlehem, T., Abhayapala, T.D.: Analysis of 2D sound reproduction with fixed-directivity loudspeakers. In: 2012 IEEE International Conference on Acoustics, Speech and Signal Processing, pp. 377–380. IEEE Press, Kyoto (2012). https://doi.org/10.1109/ICASSP.2012.6287895
14. Allen, J.B., Berkley, D.A.: Image method for efficiently simulating small-room acoustics. J. Acoust. Soc. Am. **65**(4), 943–950 (1979). https://doi.org/10.1121/1.2003643

Recognition of Underwater Acoustic Target Using Sub-pretrained Convolutional Neural Networks

Andi Pan[1], Xi Chen[1], and Wei Li[1,2](✉)

[1] School of Computer Science, Fudan University, Shanghai 201203, China
weili-fudan@fudan.edu.cn

[2] Shanghai Key Laboratory of Intelligent Information Processing, Fudan University, Shanghai 201203, China

Abstract. Underwater acoustic target recognition is the task of classifying targets using ship-radiated noise in the marine environment. It is incredibly hard and complex for the complexity of the marine environment. Before the popularization of deep learning, conventional target recognition methods are mainly based on the audio time-frequency domain analysis. Different targets have obvious variation in some frequency bands, which leads to the inability of traditional methods to make full use of spectral information. In order to extremely extract the information in each frequency bands, this paper proposes a novel Sub-pretrained CNNs. For each frequency band in the spectrogram, a CNN classifier is trained on the training set. Finally, the features extracted by each CNN and the position embedding of the frequency band are concatenated as the input of the global classifier. Compare with state of the art method, the paper achieves better performance. As the experimental results show, the identification performance of UATR can be enhanced by the Sub-pre-trained CNNs method.

Keywords: Convolutional Neural Networks · Audio classification · Underwater acoustic target recognition · Pre-training

1 Introduction

Underwater acoustic target recognition is the task of classifying targets using ship-radiated noise in the marine environment. It is widely used for marine exploration, marine biological surveys, and other research activities. It is incredibly hard and complex for the complexity of the marine environment and the diversity of underwater acoustic targets [1,2].

At present, various UATR methods based on machine learning have been put forward. Commonly, we separate these methods into two kinds: approaches based on artificial feature design and approaches based on automatic feature extraction [1–4]. In general, the most effective method of UATR is based on the characteristics of domain knowledge design, which heavily depends on the

X. Shao et al. (Eds.): CSMT 2020, LNEE 761, pp. 113–123, 2021.
https://doi.org/10.1007/978-981-16-1649-5_10

statistical model [1–3]. MFCC is a widely adopted feature in UATR and speech recognition [3–13]. Nevertheless, the optimal feature of the acoustic target can not be represented by MFCC [8]. To solve the shortcomings of MFCC, other features have been presented. The GFCC was introduced into UATR by Lian [9]. The crux in the process is how to extract the features of underwater acoustic targets.

In recent years, as the solution based on deep learning has made great successes in the field of speech recognition and image classification, people have carried out in-depth research on improving the ability of underwater acoustic target recognition. [14–19] in these studies, the solution based on deep learning shows a strong ability to feature extraction. Compared with the shallow neural network, the deep neural network can extract more abstract and higher-level features from big data [21]. As one of the methods based on deep structure, Deep Boltzmann Machine has better performance in learning and extracting the features of ship radiated noise. Additionally, CNNs [23] are widely used in UATR because of its advantage in processing images [24]. In [25], Yang et al. used ADCNN to simulate the auditory system. Deep learning based methods can extract more information compared with hand-engineering methods.

This paper proposed an Sub-pretrained CNNs based method which combines multi-dimensional feature extracted by CNNs with the position encoding, as the input of the global classifier using fully connected DNN. Firstly, we translate original signals to time-frequency presentations as images. Then, we transform the position of bands in the spectrogram to position encoding. After we concat position encoding and multi-dimensional feature extracted by CNNs, global classifier can recognition underwater targets using the input.

In the second section, the UATR method presented is introduced detailedly. The specific content of the experimental setting is introduced in The third part. The experimental results are addressed in the fourth section. The fifth part summarizes the full paper.

2 Proposed Method

2.1 Framework

For most UATR methods, the process can be divided into feature extraction stage and learning stage. The purpose of CNN is to adopt a deep hidden structure in the perceived signal to produce a great feature presentation. The process of the presented approach for UATR is presented in Fig. 1. As preprocess, we practice STFT to get time-frequency representations of the original signals. Next, we simply utilize each band of time-frequency representation to train each CNN model in the training dataset and train some sub-pre-trained CNNs. The outputs of the last layers of these CNNs can be considered as presentation of the band. Then, we concat vectors as just one vector. Finally, we take the vector as the input of global classifier, which will recognize the target.

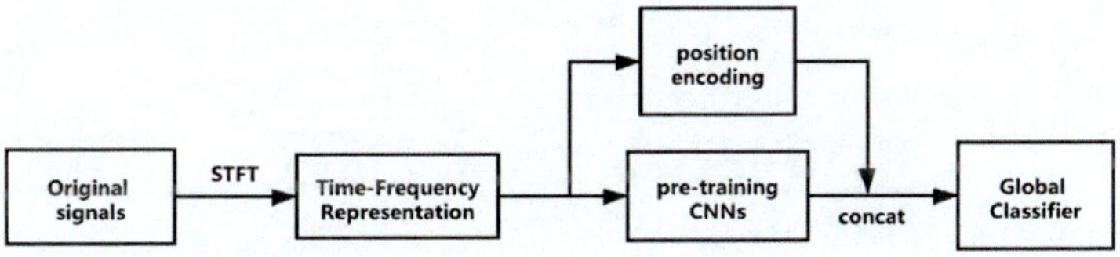

Fig. 1. The process of the presented UATR

2.2 Sub-pretrained CNNs

Spectrograms are 2D representations like an image comprising time and frequency dimensions, although very distinct from the original images. There exists an obvious diversification during the frequency dimension. As shown in Fig. 2, in the spectrograms obtained, we observed a clear variation of the magnitude of different frequency bands, particularly specific to every kind of target. For instance, the "B" class owns more extra power in higher frequency bins; the "C" class has more energy in mid-frequency bins and less energy in higher frequency bins; for "E" class Background noise recordings, energy is well-distributed in frequency bands. We utilize these observations to put forward Sub-pre-trained CNNs, which is talked about in the accompanying.

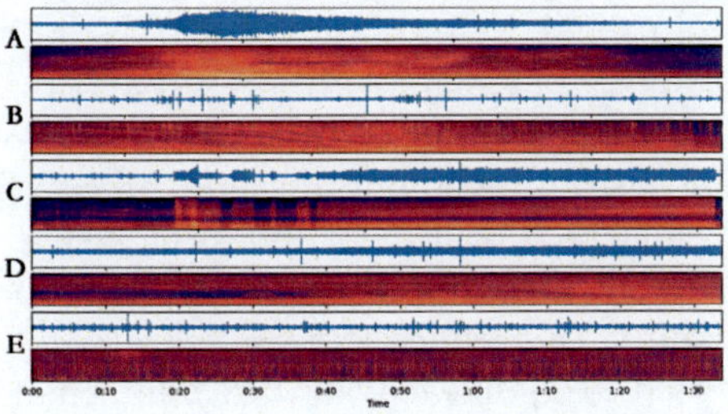

Fig. 2. Time-frequency presentation

To extremely extract the information in each frequency band and fully take advantage of variation of the magnitude of different frequency bands, we propose the Sub-pretrained CNNs method. The process of this method can be illustrated in Fig. 3. Firstly, we extract the spectrogram for the N samples and perform normalization. Then we split the spectrogram into several bands. It takes spectrogram to F*T dimension, bands size is the number of bands. These bands are independently inputted into 2 conv-layers. Kernel-size is set (5, 5), which has large receptive field. After conv-layer, sigmoid activation and max-pooling follow. Then, we flatten the output of CNNs and concat these vectors as just one vector. Finally, to capture the global relations between frequency bands, we use MLP as classifier to classify the input using diversified information.

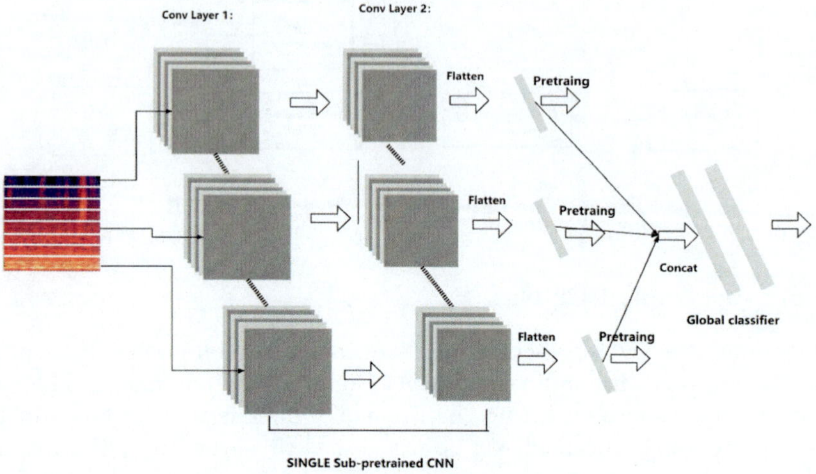

Fig. 3. Sub-pretrained CNNs

2.3 Position Encoding

Position and order of bands are the essential parts of any spectrogram. They define the high and low frequency and thus the actual characteristics of an acoustic target. Convolutional Neural Networks (CNNs) rarely take the order of bands into account. They parse a spectrogram band by band in a sequential manner. This will integrate the bands' order.

This paper use the position encoding method proposed in Transfomer [26], which is a simple yet efficient tool. Firstly, it is not just a number. Instead, it's a d-dimensional vector that incorporates information about a specific position in a spectrogram. Secondly, this vector is not integrated into the classifier itself. Instead, this vector is used to equip each word with information about its position in a spectrogram. Basically, we enhance the classifier's input to inject the order of bands.

$$PE(pos) = sin(\frac{pos}{length}) \tag{1}$$

2.4 Classifier

We test SVM, Decision Tree and MLP as classifier. The performance of classifiers are shown in Sect. 4. The methods are implemented with scikit tools. The principles of these algorithms are introduced as follows.

Support Vector Machine. SVM [27] is a very classical and commonly used model. Because it has very good classification ability and strong interpretability, it has a good effect on small samples. For linearly separable data, linear support vector machine strives to find a segmentation line to maximize the distance between positive and negative samples. When the data is approximately separable but not completely separable and not completely separable, there are a small number of abnormal samples. Using soft margin maximization, we can fit a classifier that basically separates the samples but can not completely separate them. When the data set can not be divided by the interval represented by the linear function, someone put forward the kernel function to convert the original data space where the training set samples exist toward a higher dimensional feature space, formerly the data set converts separable. In order to train a nonlinear classifier, the principle is shown in the figure. The common kernel functions are Gaussian kernel and so on.

Decision Tree. Decision Tree [28] is a model that accords with human judgment intuition and has strong explanation. After abstraction, the decision tree model is generally more like a tree, so it is named decision tree. As shown in Fig. 4, the segmentation part of the branches in this structure is to select a feature in the sample features to segment the data set. The decision book belongs to supervised learning.

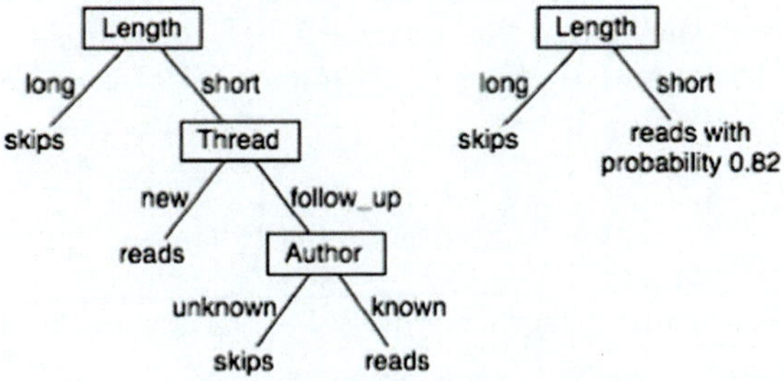

Fig. 4. Two decision trees.

Multi-layer Perceptron. Perceptron (Perceptron) is the origin of deep learning. Through the weight w and the offset term b, it can map a multi-dimensional input X to a binary value, through which a simple binary classification can be achieved. Multilayer perceptrons are in the form of multiple functions. As shown in Fig. 5, the multilayer perceptron is the superimposed multiple function of the function represented by the perceptron, which is divided into input, output, concealment and multiple perceptrons according to function and position. At the same time, if each unit of the multilayer perceptron is linear, then any multilayer perceptron can be equivalent to a single layer perceptron. Therefore, the multilayer perceptron is essentially the superposition of multiple nonlinear functions. Finally, the model is used to measure the fitting degree of the training set, and the variables in the model are taken as the loss function of the parameters. Through the back propagation algorithm, a multi-layer perceptron can be fitted on the training set.

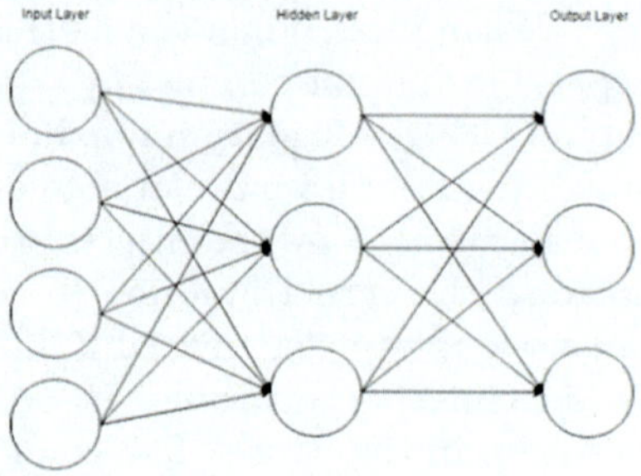

Fig. 5. Structure of Multi-layer Perceptron.

3 Experiments Setup

3.1 Experimental Datasets

The ship target dataset used in this paper is the ShipsEar [30] dataset recorded in different regions of the Spanish coast from 2012 to 2013. The dataset has a total of 90 records of 11 ship types within 15 s to 10 min. According to the original labels of the dataset, they can be merged into 4 large groups in accordance with the type of ship. Class and E class: background noise recordings, The detailed division is shown in Table 1 below:

Table 1. ShipEar dataset details.

A	Fishing boats.Trawlers.Mussel boats.Tugboats.Drafgers
B	Motorboats.Pilot boats.Sailboats
C	Passenger ferries
D	Ocean liner.Ro-Ro vessels
E	Background noise recordings

3.2 Training Setup

We choose 52,734 Hz as the target audio signal sampling rate, and a 90 ms Hamming window as windowing function with a 50% overlap is used. The output Mel spectrum is stored in a $3 \times 224 \times 224$ image format for subsequent operations. In addition, we downsampled the experimental audio data. The window length is 25 ms, the overlap length (Hop size) is 10 ms, the output spectrum is 96×64, and the embedding code size is 128.

We implement Sub-pretrained CNNs in Pytorch. Most experiments have been carried out with sklearn [31].

3.3 Evaluation Indexes

We compare the predicted results of the model with the labels to obtain the number of TP, FP, TN, and FN in the evaluation. And for each experimental result, the accuracy rate, recall rate, and F1 function are calculated separately to measure the experimental results comprehensively and accurately. These indicators can be expressed by the following formula:

$$Accuracy = \frac{TP}{TP + FP + TN + FN} \tag{2}$$

4 Experiments Results

4.1 Bands Size Setting

To find optimal Bands size, the experiment was designed. We set the optional values of band size to 10, 20, 30, 40. In the contrast experiment, the classification accuracy reaches the highest when band size equal to 20. Therefor, we set the band size to 20 in the following experiments.

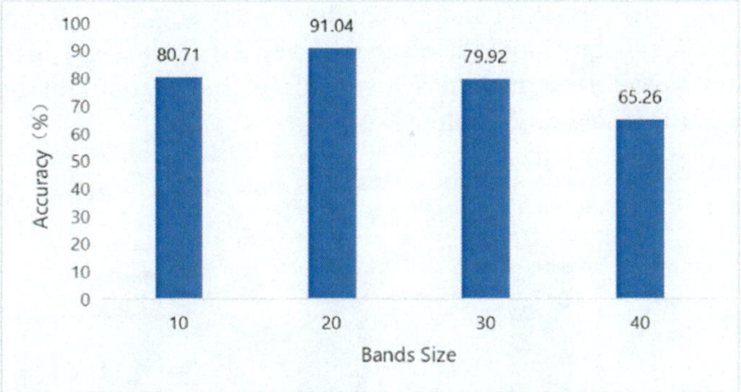

Fig. 6. The recognition accuracy with different bands size

4.2 Evaluation of Position Encoding

To illustrate the importance of position encoding, classification performance of MLP with encoding and without are measured using the classification accuracy. The comparison between MLP with position encoding and MLP without position is shown in Fig. 7. It is clear that position encoding can introduce more structure information in spectrogram, which contributes to improving the performance.

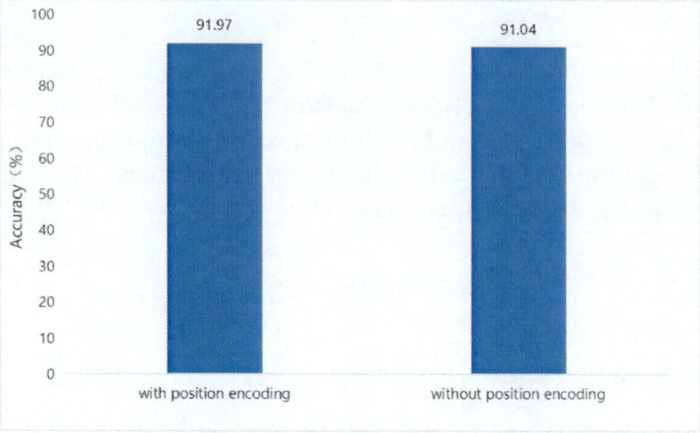

Fig. 7. The comparison between MLP with position encoding and MLP without position encoding

4.3 A Comparison of Three Kind of Classifiers

To find the optimal classifier, we compare three kinds of classifiers. As illustrated in Fig. 8, MLP classifier has the highest accuracy over Decision Tree and SVM. In contrast recognition, it is clear that the MLP classifier is more suitable for underwater target recognition. We speculate that this might be due to the advantage of MLP in classification.

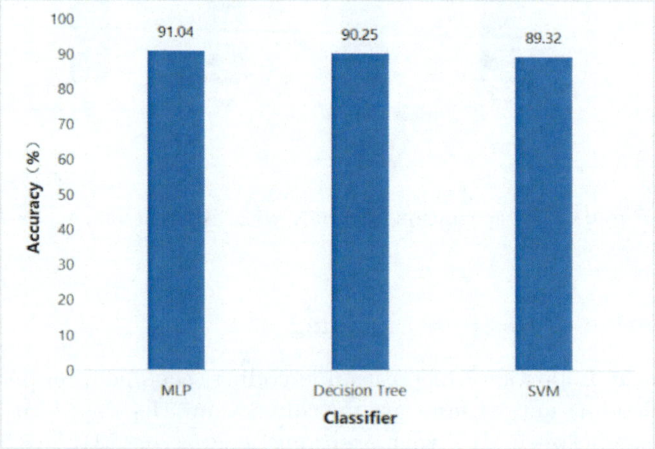

Fig. 8. The recognition accuracy with different bands size

4.4 Evaluation of Sub-pretrained CNNs with Position Encoding

Considering that methods with position encoding can achieve better performance than without, we train a Sub-pretrained CNNs with position encoding and using MLP as the classifier. As a result, accuracy is 91.97%. This best result shows in the confusion matrix. Table 3 shows the confusion matrix of the proposed UATR methods obtained from testing data. Compare with state of the art method, the paper achieves better performance (Table 2).

Table 2. Comparison of performance between Pretrained CNNs and DBM based.

Method	Accuracy
Sub-pretrained CNNs with position encoding	91.97%
DBM [22]	90.70%
VGGISH [32]	89.22%

Table 3. Confusion matrix of the proposed model.

True predicted	A	B	C	D	E
A	0.92	0.01	0.02	0.01	0.00
B	0.01	0.85	0.03	0.02	0.00
C	0.02	0.00	0.93	0.01	0.00
D	0.01	0.03	0.00	0.92	0.01
E	0.00	0.01	0.00	0.00	0.98

5 Conclusions

In the work, a new UATR algorithm based on regional pre-training convolution neural network is introduced, in order to fully extract the information contained in different frequency bands in the spectrum. The output of the last hidden layer of each sub-network is spliced and connected with the position vector as the input of the total classifier, and then the general classifier is trained. Compare with state of the normal training convolution neural network model, the proposed UATR algorithm achieves better performance, the sub-pre-trained CNN is introduced to learn more information, and the classification accuracy is 91.97%. This method proposes an innovative model training method, which can be effectively applied to UATR tasks, also give inspiration to other similar tasks.

Acknowledgement. This work was supported by National Key R&D Program of China (2019YFC1711800) and NSFC (61671156).

References

1. Yang, H., Shen, S., Yao, X., Sheng, M., Wang, C.: Competitive deep-belief networks for underwater acoustic target recognition. Sensors **18**, 952 (2018)
2. Wang, X., Jiao, J., Yin, J., Zhao, W., Han, X., Sun, B.: Underwater sonar image classification using adaptive weights convolutional neural network. Appl. Acoust. **146**, 145–154 (2018)
3. Wang, W., Li, S., Yang, J., Liu, Z., Zhou, W.: Feature extraction of underwater target in auditory sensation area based on MFCC. In: 2016 IEEE/OES China Ocean Acoustics (COA), pp. 1–6. IEEE (2016)
4. Yue, H., Zhang, L., Wang, D., Wang, Y., Lu, Z.: The classification of underwater acoustic targets based on deep learning methods. In: 2017 2nd International Conference on Control, Automation and Artificial Intelligence (CAAI 2017), pp. 526–529. Atlantis Press (2017)
5. Lu, Z., Zhang, X., Zhu, J.: Feature extraction of ship-radiated noise based on mel frequency cepstrum coefficients. Ship Sci. Technol. **26**(2), 51–54 (2004)
6. Ke, X., Yuan, F., Cheng, E.: Underwater acoustic target recognition based on supervised feature-separation algorithm. Sensors **18**(12), 4318 (2018)
7. Zhang, L., Wu, D., Han, X., Zhu, Z.: Feature extraction of underwater target signal using mel frequency cepstrum coefficients based on acoustic vector sensor. J. Sens. **2016**, 1–11 (2016)
8. Sharma, R., Vignolo, L., Schlotthauer, G., Colominas, M., Rufiner, H.L., Prasanna, S.: Empirical mode decomposition for adaptive AM-FM analysis of speech: a review. Speech Commun. **88**, 39–64 (2017)
9. Lian, Z., Xu, K., Wan, J., Li, G.: Underwater acoustic target classification based on modified GFCC features. In: Proceedings of the IEEE 2nd Advanced Information Technology, Electronic and Automation Control Conference (IAEAC), Chongqing, China, 25–26 March 2017, pp. 258–262 (2017)
10. Lim, T., Bae, K., Hwang, C., Lee, H.: Underwater transient signal classification using binary pattern image of MFCC and neural network. IEICE Trans. Fundam. Electron. Commun. Comput. Sci. **E91A**, 772–774 (2008)
11. Jankowski Jr., C., Quatieri, T., Reynolds, D.: Measuring fine structure in speech: Application to speaker identification. In: Proceedings of the International Conference on Acoustics, Speech, and Signal Processing, Detroit, MI, USA, 9–12 May 1995, pp. 325–328. IEEE, Piscataway (1995)
12. Guo, Y., Gas, B.: Underwater transient and non transient signals classification using predictive neural networks. In: Proceedings of the 2009 IEEE/RSJ International Conference on Intelligent Robots and Systems, St. Louis, MO, USA, 10–15 October 2009, pp. 2283–2288 (2009)
13. Hu, G., Wang, K., Peng, Y., Qiu, M., Shi, J., Liu, L.: Deep learning methods for underwater target feature extraction and recognition. Comput. Intell. Neurosci. **2018**, 1214301 (2018)
14. Jiang, Y., Wang, D.L., Liu, R.S., Feng, Z.M.: Binaural classification for reverberant speech segregation using deep neural networks. IEEE/ACM Trans. Audio Speech Lang. Process. **22**(12), 2112–2121 (2014)
15. Lee, H., Yan, L., Pham, P., Ng, A.Y.: Unsupervised feature learning for audio classification using convolutional deep belief networks. In: Proceedings of the 23rd Annual Conference on Neural Information Processing Systems (NIPS 2009), vol. 9, pp. 1096–1104, December 2009

16. Jaitly, N., Hinton, G.: Learning a better representation of speech soundwaves using restricted Boltzmann machines. In: Proceedings of the 36th IEEE International Conference on Acoustics, Speech, and Signal Processing, (ICASSP 2011), pp. 5884–5887, May 2011
17. Palaz, D., Collobert, R., Magimai-Doss, M.: Estimating phoneme class conditional probabilities from raw speech signal using convolutional neural networks. In: Proceedings of the 14th Annual Conference of the International Speech Communication Association, INTERSPEECH 2013, pp. 1766–1770, August 2013
18. Huang, G., Huang, G.-B., Song, S., You, K.: Trends in extreme learning machines: a review. Neural Netw. **61**, 32–48 (2015)
19. Abdel-Hamid, O., Mohamed, A.-R., Jiang, H., Deng, L., Penn, G., Yu, D.: Convolutional neural networks for speech recognition. IEEE Trans. Audio Speech Lang. Process. **22**(10), 1533–1545 (2014)
20. Bisot, V., Serizel, R., Essid, S., et al.: Acoustic scene classification with matrix factorization for unsupervised feature learning. In: IEEE International Conference on Acoustics. IEEE (2016)
21. Kamal, S., Mohammed, S.K., Pillai, P.R.S., Supriya, M.H.: Deep learning architectures for underwater target recognition. In: Proceedings of Ocean Electronics (SYMPOL), October 2013, pp. 48–54 (2013)
22. Hinton, G., et al.: Deep neural networks for acoustic modeling in speech recognition: the shared views of four research groups. IEEE Signal Process. Mag. **29**(6), 82–97 (2012)
23. Deng, L., Abdel-Hamid, O., Yu, D.: A deep convolutional neural network using heterogeneous pooling for trading acoustic invariance with phonetic confusion. In: Proceedings of IEEE International Conference on Acoustics, Speech, and Signal Processing (ICASSP), May 2013, pp. 6669–6673 (2013)
24. Swietojanski, P., Ghoshal, A., Renals, S.: Convolutional neural networks for distant speech recognition. IEEE Signal Process. Lett. **21**(9), 1120–1124 (2014)
25. Yang, H., Li, J., Shen, S., Xu, G.: A deep convolutional neural network inspired by auditory perception for underwater acoustic target recognition. Sensors **19**, 1104 (2019)
26. Ott, M., Edunov, S., Baevski, A., et al.: FAIRSEQ: a Fast, extensible toolkit for sequence modeling. In: Proceedings of the 2019 Conference of the North (2019)
27. Suykens, J.A.K., Vandewalle, J.: Least squares support vector machine classifiers. Neural Process. Lett. **9**(3), 293–300 (1999)
28. Zhou, B., Cao, C., Li, C., et al.: Hybrid islanding detection method based on decision tree and positive feedback for distributed generations. IET Gen. Transm. Distrib. **9**, 1819–1825 (2015)
29. Yue, H., Zhang, L., Wang, D., Wang, Y., Lu, Z.: The classification of underwater acoustic targets based on deep learning methods. Adv. Intell. Syst. Res. **134**, 526–529 (2017)
30. Santos-Domínguez, D., Torres-Guijarro, S., Cardenal-López, A., et al.: Shipsear: an underwater vessel noise database. Appl. Acoust. **113**, 64–69 (2016)
31. Pedregosa, F., Varoquaux, G., Gramfort, A., et al.: Scikit-learn: machine learning in python. J. Mach. Learn. Res. **12**, 2825–2830 (2011)
32. Deng, J., Pan, A., Xiao, C., Chen, S.: Transfer learning for acoustic target recognition. Comput. Syst. Appl. **29**(10), 255–261 (2020)

Two-Stage Classification Learning for Open Set Acoustic Scene Classification

Chunxia Ren[1] and Shengchen Li[2(✉)]

[1] Beijing University of Posts and Telecommunications, No. 10 Xitucheng Road, Haidian District, Beijing, China
chunxiaren@bupt.edu.cn

[2] Department of Intelligent Science, School of Advanced Technology, Xi'an Jiaotong-Liverpool University, 111 Ren'ai Road, Suzhou Industrial Park, Suzhou 215123, Jiangsu Province, P. R. China
shengchen.li@xjtlu.edu.cn

Abstract. Most of the research on acoustic scene classification (ASC) focuses on classification problem with only known scene classes. In practice, scene classification problem to be solved generally is based on an open set, which contains unknown scenes. This paper proposes a two-stage method that solves the open set problem on ASC. The proposed system decomposes open set ASC problem into two stages. To mitigate the impact of unknown scenes on the subsequent recognition process of known scenes, the first stage is to identify unknown scenes. The second stage classifies defined acoustic scenes. In this case, the threshold selection strategy we proposed further sorts out unknown scenes that were not identified in the previous stage. Experiments show that the method proposed in this paper can effectively identify unknown scenes and classify known scenes, by segmenting the open set acoustic scene classification task and selecting an appropriate judgment threshold. On the development dataset released by DCASE Challenge 2019 Task 1C, the model proposed outperforms the first place.

Keywords: Acoustic scene classification · Open set · Two-stage classification · Threshold selection strategy

1 Introduction

As an environmental identification problem, acoustic scene classification (ASC) attracts growing attention [2]. ASC processes the audio signal and then extracts feature information, and the scene is identified by event or semantic information contained in feature representation [16]. ASC is widely used in smart wearable devices, robots, home surveillance and security systems, environmental noise monitoring.

The challenge of Detection and Classification of Acoustic Scenes and Events (DCASE) [15] provides a series of the open-source database and evaluation methods which develop ASC. In recent years, Convolutional Neural Networks (CNNs), Recurrent Neural Networks (RNNs) and Convolutional Recurrent Neural Networks (CRNNs) are recognised as effective models for ASC problems and

X. Shao et al. (Eds.): CSMT 2020, LNEE 761, pp. 124–133, 2021.
https://doi.org/10.1007/978-981-16-1649-5_11

are generally superior to traditional machine learning methods in performance [1,17,22]. However, the research of ASC is mainly focused on the closed set, that is, the scene classes used in the testing phase and the training phase are same. In practical scene analysis application, undefined scene classes other than limited known scene classes are often encountered. Thus, open set recognition task [4–6] which needs to additionally identify the undefined scenes as an unknown class is more useful despite higher complexity.

This paper focuses on solving open set ASC problem. The significant differences in data composition between open set and closed set make traditional ASC models no longer applicable to open set ASC. To the best of our knowledge, the research based on open set classification tasks is mainly based on one-stage classification methods [7,8,18]. Daniele et al. [3] firstly proposed a solution to the open set problem in the ASC field. They not only use Support Vector Data Description (SVDD) classifier to learn a hypersphere from known scenes to distinguish unknown class but also introduce a new protocol and indicator for evaluating the open set ASC task. The introduction of this question has attracted some scholars to study.

The DCASE Challenge 2019 Task 1C further facilitates extensive research in open set ASC. These solutions proposed by Wilkinghoff et al. [20] and Lehner et al. [13] classify known classes and separate unknown classes only by learning known classes in a single classification system; the difference is that the former used Deep Convolutional Auto-Encoders (DCAEs) as classification model and the latter used the improved ResNet variant [11,12] as the classifier. These one-stage classification methods [13,20] in which unknown classes do not participate in training phase pay more attention to the inter-class differences of known scenes but are not necessarily useful for separating unknown scene from known scenes. A one-stage classification method proposed by Zhu et al. [23] is to put the unknown class into training phase and designs an $K+1$ classifier that treats the unknown class like K known classes. This method uses CRNN-Attention mechanism model [19,21] as the classifier and achieves the first place of the DCASE Challenge 2019 Task 1C. There is a problem with this method. Although the unknown classes participate in training process, the operation where unknown classes are unreasonably regarded as a known scene is likely to ignore the difference between the unknown class and the entire set of known classes in the distribution of the feature space.

To avoid the problems of the two types of methods mentioned above [14, 20,23], this paper proposes a solution for open set ASC. Unknown classes are no longer considered to be an equal role for K known classes in this paper. Consequently, this paper designs a two-stage classification learning system for open set ASC to better solve the open set classification problem. The first stage is used to distinguish unknown classes from the entire set of known classes, reducing the impact on the next stage. The second stage is used to further divide the known classes into defined scene labels and separate the remaining unknown classes which not identified during first one, as well as we proposed a threshold selection strategy to assist in identification of unknown classes. Experiments show that our proposed two-stage method which identifies unknown classes and classifies known classes more precisely than traditional one-stage methods is an effective open set ASC solution.

The remainder of this paper is organized as follows: Sect. 2 describes the two-stage classification method for open set ASC presented in this paper; Sect. 3 introduces experimental setup and results in analysis, and Sect. 4 summarizes current works and discusses future research directions.

2 Proposed Two-Stage Classification Model

There are two problems to be solved in the open set ASC task, one is to identify unknown scene, and the other is to classify known scenes. Therefore, in this section, a two-stage classification learning model for open set ASC is proposed. The system divided the open set ASC question into two parts and resolves them in two stages. This section describes in detail how the two-stage ASC system (as shown in Fig. 1) proposed for the open set completes the classification task.

2.1 Two-Stage Classification Model—The First Stage

The two-stage classification system for open set ASC is proposed in this paper (as shown in Fig. 1). The main role of the first stage is to separate the same or similar unknown class encountered in training phase, these separated unknown class no longer participates in the second stage testing phase, reducing the impact on the second stage classification.

There are two reasons for unknown class to participate in the first stage training phase. One is to make the model not overemphasize discriminative features

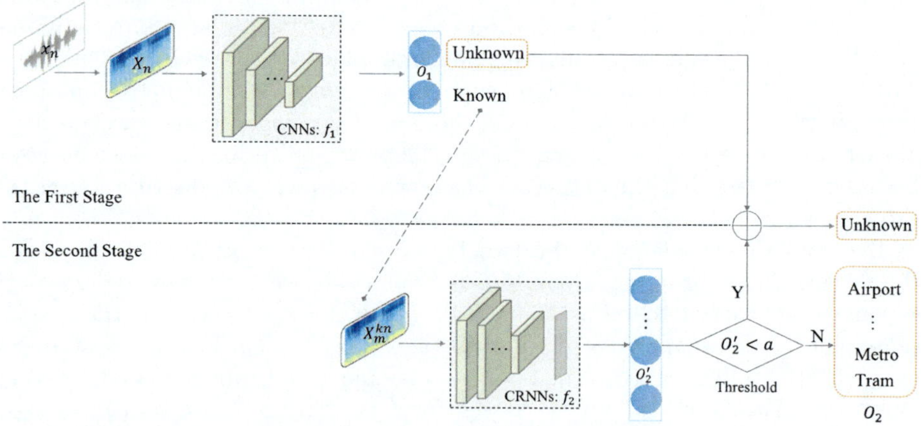

Fig. 1. A two-stage open set ASC system proposed is consist of the first stage and the second stage classification models. The output O_2 is the sum of K known classes obtained by the second stage classification model and unknown class obtained by the first stage and the second stage classification models.

of the known classes by adding the unknown class samples; the other is to fully consider the situation that may occur during testing phase because it does not know in advance whether the sub-scenes contained in the unknown class of the testing phase have already been encountered in the training phase.

To classify the scenes into the known class or unknown class, CNNs which performed well in ASC task is used as the first stage classification model [17,22]. After decomposing the original open set ASC task into two problems solved in two stages respectively, the task complexity of the first stage is significantly reduced, then the requirement for model complexity is also reduced. Thus, the first stage task can be accomplished using a CNN classification model with shallow structure. The CNNs proposed in this paper is composed of four convolution layers followed by maxpooling layer.

In the first stage (Fig. 1 upper part), the features X_n which represented by the log-mel spectrogram of the scene audio signal x_n is taken as input, where n represents the index of audio. The advanced feature representation of the original input X_n is extracted by the shallow CNNs with four layers. The distinctive information of the features of the known class and the unknown class that appeared in the training phase is learned and used as a classification basis. The global average pooling (GAP) is used to convert the feature map of the last layer of CNNs into feature points by averaging pooling. Thus, feature points with significant visibility in CNNs are reserved by the GAP. The output p_n of the neural network is a predicted probability that indicates whether the sample belongs to the unknown class. To this end, the model is optimized by updating the weights during backpropagation and minimizing the binary cross-entropy loss:

$$l = -\sum_{n=1}^{N}((y_n \log p_n) + (1 - y_n)\log(1 - p_n)) \tag{1}$$

where N is the number of samples in training phase, y_n represents the estimated label of the nth sample. Finally, the samples of the first stage are expected to be classified as known class or unknown class.

2.2 Two-Stage Classification Model—The Second Stage

Compared with the first stage, the second stage needs to detailly classify the complex known scenes into K defined classes. From the perspective of task complexity, the second stage is more complicated, which may result in shallow CNNs does not necessarily complete the task well. Since CRNNs was proposed by [1], there has been a lot of work to prove its excellent performance on ASC. Therefore, we use it as the classification model in the second stage [21].

As shown in the lower part of Fig. 1, the features $X^k_m n$ of the known scenes $x^k_m n$ is used as input to CRNNs, where m is denoted as the index of audio during the second stage. Among CRNNs, CNNs which acts as advanced features extractor passed the abstracted advanced feature information into bi-directional RNN (Bi-RNN). The information in features that helps to classify scenes is not only independent, but it is also sometimes related to its occurrence time. Therefore, considering the need to maintain the temporal resolution of the sequence generated by Bi-RNN, the pooling operation only occurs on the frequency axis. In this way, Bi-RNN learned the contextual timing relationship of the feature and encode it. Bi-RNN regarded as another advanced feature extractor, but different from CNNs mode.

In the first stage, some unknown scenes used for the testing phase that differs greatly from the trained unknown scenes in the feature space may be missed. Since the classifier in the second stage is designed for known scenes, it could be assumed that the probability of unknown scenes is relatively low. Therefore, a judgment threshold h is needed to determine the scene with an output probability below h as an unknown class. To make the threshold at this stage divide the known and unknown classes more scientifically, we propose a threshold selection strategy. If M_{uk} which is the number of unknown samples in the testing phase is known, the choice of threshold hs should make predicted probability of at least M_{uk} testing samples lower than h. When M_{uk} is unknown, but the relationship between the accuracy of the unknown classes and the accuracy of the system is known as:

$$ACC = (1 - \beta) * ACC_{kn} + \beta * ACC_{uk} \tag{2}$$

Then the value of the threshold should result in the predicted probability of $\beta * N_t$ samples being lower than h, where β is a weight coefficient which less than 1 but over 0 and N_t is the number of testing samples.

The weighted average operation proposed by [21] is used in the second stage to obtain a suitable probability output so that the selected threshold h separates the unknown class which is not identified in the first stage. The weighted average operation is following,

$$O_2' = \frac{\sum_{t=0}^{T-1} O'(t)}{\sum_{t=0}^{T-1} Z_{soft}(t)} \tag{3}$$

where

$$O'(t) = Z_{soft}(t) \odot Z_{sigm}(t) \tag{4}$$

T is the frame-level resolution and O' is the element-wise multiplication of the outputs of two fully connected layers whose activation function is softmax Z_{soft} and sigmoid Z_{sigm}.

Then, the output corresponding to the first stage is $O_1\{O_1^{kn}; O_1^{uk}\}$, the output of the second stage is O_2', and the further output after the threshold a judgment is $O_2\{O_2^{kn}; O_2^{uk}\}$. The output O composition of our proposed system should be the combination of the sum of the unknown class identified in the first stage and the second stage, and the classification results of the K known classes in the second stage as following,

$$Output\colon O = (O_2^{kn}; O_1^{uk} \bigcup O_2^{uk}) \tag{5}$$

3 Experiments

3.1 Dataset and Experimental Setup

This paper verified the two-stage classification system for open set ASC presented on the development dataset published by the Task 1C of DCASE 2019 Challenge. The dataset contains known scenes and unknown scenes; the former is 10 scenes recorded in 10 different European cities, each recording approximately 1440 audio samples; the latter consists of 4 different sub-scenes, the number of audio samples recorded in each scene is about 480. The duration of the audio samples is 10 s.

The ratio of the training set and the testing set is 3:1. 10 visible sub-scenes of known class appear in both training and testing sets. There are two possible situations where invisible sub-scenes of unknown class in the testing set may be completely different from sub-scenes of unknown class in the training set or maybe partial duplication. To demonstrate the effectiveness of the proposed system, we do a set of comparative experiments. The parameter setups in the experiment are as follows.

Figure 2 shows the composition of the classification models in the first and second stages. Log-mel spectrogram is used as the features of audio samples, with 640 frames per chunk by 128 mel bins, and then each chunk is evenly divided into 5 segments, each segment has 128 frames. Batch normalization [9] is applied after each convolutional layer. During the experiment, dropout was added to avoid over-fitting of the proposed model, the judgment threshold was chosen to be 0.2 by threshold selection strategy, and the Adam optimizer with learning rate which fixed at 0.001 is used.

3.2 Results and Analysis

The number of correctly classified audio samples in the total number of audio samples called classification accuracy is used as the score of the open set ASC. Accuracy is calculated as the weighted average of the known classes and unknown class, as shown below:

$$ACC_{weighted} = 0.5 * ACC_{kn} + 0.5 * ACC_{uk} \tag{6}$$

where known classes accuracy ACC_{kn} is the average of the class-wise accuracy.

In Table 1, this experiment compares the proposed model with two typical one-stage models on the development dataset divided by the Task 1C of DCASE 2019 Challenge. These two typical models are the Baseline and the best model [23] published on DCASE 2019 Challenge, respectively. Among them, the 10 classification model based on CNNs is adopted by the Baseline, with 0.5 as the judgment threshold; the 11 classifications based on CRNN-Attention model [19] is adopted by the model [23]. And our proposed two-stage classification learning system achieves better results with nearly 5% improvement over the best model by identifying unknown classes in the first stage, classifying known classes

First stage (CNNs)	Second stage (CRNNs)
Log-mel spectrogram 128 frames * 128 mel bins	Log-mel spectrogram 128 frames * 128 mel bins
$\left[\begin{pmatrix} 3*3\ @64 \\ BN, ReLU \end{pmatrix}, P(1,2)\right] * 2$	$\left[\begin{pmatrix} 3*3\ @128 \\ BN, ReLU \end{pmatrix}, P(1,2)\right] * 5$
$\left[\begin{pmatrix} 3*3\ @128 \\ BN, ReLU \end{pmatrix}, P(1,2)\right] * 2$	$\left[\begin{pmatrix} 3*3\ @64 \\ BN, ReLU \end{pmatrix}, P(1,4)\right]$
Global average pooling	Bi-GRU @128
	Weighted average

Fig. 2. The models of the first and second stages. 'P' represents pooling, 'BN' is Batch Normalization.

unknown classes that are difficult in the previous stage in the second stage. Both the average accuracy of known classes and the accuracy of the unknown class are higher than the one-stage classification methods, which proves that our proposed two-stage method is more reasonable and has better performance for unknown classes recognition and known classes classification. To further compare the difference in the class-wise accuracy between the system proposed in this paper and the other two systems, Table 1 shows the comparison of the class-wise average accuracy of scene classes on these three models. It can be seen that the proposed system has the highest accuracy in multiple scene classes, but it does not perform well in individual classes such as "Airport", "Street_pedestrian", and "Tram". One possible reason is that these low-accuracy scenes are similar to other scenes in the feature space, causing the system to misjudge.

Compared with several other models that use a fixed empirical threshold, the model we proposed verifies the rationality of the threshold selection strategy. The model [10] and Baseline in Table 2 utilize a traditional threshold of 0.5 to identify samples with prediction probability lower than 0.5 as "Unknown". This choice leads to the randomness of results, and it is difficult to ensure that 0.5 is the appropriate probability boundary between the known classes and the unknown classes. These models do not take into account the probability of prediction and data composition together. The threshold of model [23] is selected as 0.4, and the same problem exists. We choose the threshold as 0.2 based on the threshold selection mechanism proposed in this paper.

Since the number of samples of the unknown class in the testing phase is 345, accounting for nearly 7.6% of the testing sample. 0.2 is selected as the threshold according to the threshold selection strategy so that the prediction

Table 1. The accuracy (%) of the corresponding model. Among them, "Known" represents the average accuracy of 10 known scene classes, "Unknown" represents the accuracy of unknown scene classes, and "Overall" is the accuracy calculated by Formula (6).

Accuracy	Model		
	Baseline	Zhu et al. [23]	Our model
Airport	44.2	65.3	41.1
Shopping_mall	50.9	26.3	**71.7**
Metro_station	41.3	42.1	**56.6**
Public_square	34.7	39.8	**45.0**
Metro	51.5	42.3	**51.7**
Tram	60.7	57.6	55.1
Street_pedestrian	47.5	37.3	46.9
Street_traffic	78.4	74.4	**80.4**
Bus	59.3	52.3	53.7
Park	74	80.8	64.8
Known	54.3	51.8	**56.7**
Unknown	43.1	75.9	**80.3**
Overall	48.7	63.9	**68.5**

Table 2. The relationship between selected threshold and accuracy (%) of the corresponding model. "Unknown" represents the accuracy of unknown scene classes, and "Overall" represents the accuracy calculated by Formula (6).

Model	Threshold	Unknown	Overall
Baseline	0.5	43.1	48.7
Kong et al. [10]	0.5	48.1	53.1
Zhu et al. [23]	0.4	75.9	63.9
Our model	**0.2**	**80.3**	**68.5**

probability of about 7.5%–8% testing samples is lower than 0.2. As is seen from the above Table 2, the method we proposed has the highest accuracy of the unknown class: 80.3%, which has an obvious advantage than other methods. The above scheme is obtained when the number of "Unknown" samples is known. We verified the rationality of the proposed threshold selection mechanism on the dataset of private Kaggle leaderboard when the number of "Unknown" samples is unknown. Therefore, according to the formula (6), we choose a threshold of 0.4, which makes about half of the testing samples' prediction probability is lower than the threshold. Under the threshold selection strategy, we achieved the best results of the private Kaggle leaderboard.

4 Conclusion

This paper proposes a two-stage classification learning solution for open set ASC, which achieves 68.5% by using the proposed model on the development dataset of the DCASE 2019 for open set ASC, which is better than the optimal performance released by DCASE Challenge 2019 Task 1C. The experiment proves that the proposed model is really useful. In the future, we will explore how to improve the accuracy of the known classes while ensuring the unknown class accuracy, and balance the accuracy of known classes and unknown class. Besides, we will also study the less dependent experience-based solutions for open set ASC, and the feature representation methods that can more clearly distinguish different scenes.

Acknowledgements. This work was supported in part by the National Natural Science Foundation of China (62001038).

References

1. Adavanne, S., Politis, A., Nikunen, J., Virtanen, T.: Sound event localization and detection of overlapping sources using convolutional recurrent neural networks. IEEE J. Sel. Top. Signal Process. **13**(1), 34–48 (2018)
2. Barchiesi, D., Giannoulis, D., Stowell, D., Plumbley, M.D.: Acoustic scene classification: classifying environments from the sounds they produce. IEEE Signal Process. Mag. **32**(3), 16–34 (2015)
3. Battaglino, D., Lepauloux, L., Evans, N.: The open-set problem in acoustic scene classification. In: 2016 IEEE International Workshop on Acoustic Signal Enhancement (IWAENC), pp. 1–5. IEEE (2016)
4. Bendale, A., Boult, T.: Towards open world recognition. In: Proceedings of the IEEE Conference on Computer Vision and Pattern Recognition, pp. 1893–1902 (2015)
5. Bendale, A., Boult, T.E.: Towards open set deep networks. In: Proceedings of the IEEE Conference on Computer Vision and Pattern Recognition, pp. 1563–1572 (2016)
6. Chen, J., Sathe, S., Aggarwal, C., Turaga, D.: Outlier detection with autoencoder ensembles. In: Proceedings of the 2017 SIAM International Conference on Data Mining, pp. 90–98. SIAM (2017)
7. Chen, Y., Zhou, X.S., Huang, T.S.: One-class SVM for learning in image retrieval. In: ICIP, vol. 1, pp. 34–37. Citeseer (2001)
8. Ding, Z., Fei, M.: An anomaly detection approach based on isolation forest algorithm for streaming data using sliding window. IFAC Proc. Vol. **46**(20), 12–17 (2013)
9. Ioffe, S., Szegedy, C.: Batch normalization: accelerating deep network training by reducing internal covariate shift. arXiv preprint arXiv:1502.03167 (2015)
10. Kong, Q., Cao, Y., Iqbal, T., Xu, Y., Wang, W., Plumbley, M.D.: Cross-task learning for audio tagging, sound event detection and spatial localization: DCASE 2019 baseline systems. arXiv preprint arXiv:1904.03476 (2019)

11. Koutini, K., Eghbal-Zadeh, H., Dorfer, M., Widmer, G.: The receptive field as a regularizer in deep convolutional neural networks for acoustic scene classification. In: 2019 27th European signal processing conference (EUSIPCO), pp. 1–5. IEEE (2019)
12. Koutini, K., Eghbal-zadeh, H., Widmer, G., Kepler, J.: CP-JKU submissions to DCASE 2019: acoustic scene classification and audio tagging with receptive-field-regularized CNNs. In: Proceedings of the Detection and Classification of Acoustic Scenes and Events Workshop (DCASE), New York, NY, USA, pp. 25–26 (2019)
13. Lehner, B., Koutini, K., Schwarzlmüller, C., Gallien, T., Widmer, G.: Acoustic scene classification with reject option based on resnets. In: Proceedings of the Detection and Classification of Acoustic Scenes and Events Workshop (DCASE), New York, NY, USA, pp. 25–26 (2019)
14. Lei, C., Wang, Z.: Multi-scale recalibrated features fusion for acoustic scene classification (2019)
15. Mesaros, A., Heittola, T., Benetos, E., Foster, P., Lagrange, M., Virtanen, T., Plumbley, M.D.: Detection and classification of acoustic scenes and events: outcome of the dcase 2016 challenge. IEEE/ACM Trans. Audio Speech Lang. Process. (TASLP) **26**(2), 379–393 (2018)
16. Phaye, S.S.R., Benetos, E., Wang, Y.: Subspectralnet–using sub-spectrogram based convolutional neural networks for acoustic scene classification. In: ICASSP 2019-2019 IEEE International Conference on Acoustics, Speech and Signal Processing (ICASSP), pp. 825–829. IEEE (2019)
17. Ren, Z., Kong, Q., Han, J., Plumbley, M.D., Schuller, B.W.: Attention-based atrous convolutional neural networks: Visualisation and understanding perspectives of acoustic scenes. In: ICASSP 2019-2019 IEEE International Conference on Acoustics, Speech and Signal Processing (ICASSP), pp. 56–60. IEEE (2019)
18. Scheirer, W.J., Jain, L.P., Boult, T.E.: Probability models for open set recognition. IEEE Trans. Pattern Anal. Mach. Intell. **36**(11), 2317–2324 (2014)
19. Wang, J., Li, S.: Self-attention mechanism based system for dcase2018 challenge task1 and task4. In: IEEE AASP Challenge on DCASE 2018 Technical Reports (2018)
20. Wilkinghoff, K., Kurth, F.: Open-set acoustic scene classification with deep convolutional autoencoders. In: Proceedings of the Detection and Classification of Acoustic Scenes and Events Workshop (DCASE) (2019)
21. Xu, Y., Kong, Q., Wang, W., Plumbley, M.D.: Large-scale weakly supervised audio classification using gated convolutional neural network. In: 2018 IEEE International Conference on Acoustics, Speech and Signal Processing (ICASSP), pp. 121–125. IEEE (2018)
22. Yang, Y., Zhang, H., Tu, W., Ai, H., Cai, L., Hu, R., Xiang, F.: Kullback–Leibler divergence frequency warping scale for acoustic scene classification using convolutional neural network. In: ICASSP 2019-2019 IEEE International Conference on Acoustics, Speech and Signal Processing (ICASSP), pp. 840–844. IEEE (2019)
23. Zhu, H., Ren, C., Wang, J., Li, S., Wang, L., Yang, L.: DCASE 2019 challenge task1 technical report. Technical report, DCASE 2019 Challenge, Technical report (2019)

An Overview of Speech Dereverberation

Yuan Li[(✉)] and Lunhui Deng

Communication University of China, Beijing 100024, China
{biliyuan,dy3000}@cuc.edu.cn

Abstract. Speech dereverberation is an important preprocessing step in speech signal processing, aims at improving the sound quality by canceling or suppressing the effect of reverb. This paper provides an overview of speech dereverberation algorithms, showing the development of speech dereverberation technology. With the categories and summaries of existing speech dereverberation algorithms, our goal is to analyze each type of algorithms' advantages and disadvantages and provide the necessary background to the readers who are going to devote themselves to making progress in this area. Finally, the overview will provide some future work directions.

Keywords: Dereverberation · Speech dereverberation · Room impulse response

1 Introduction

With the development of 5G mobile communication and the popularization of intelligent voice control wireless mobile wearable devices, the demand for long-distance high-accuracy speech recognition technology is increasing rapidly. However, reverberation is built up in the indoor environment, particularly disruptive for speech perception, causing significant performance degradation in speech recognition. The speech signal can be effectively improved after the dereverberation. For example, the signal-to-interference ratio (SIR) in preference [1] has increased by 60 dB; and the speech recognition word error rate (WER) in preference [2] is reduced from 49.2% to 9.0%.These indicate a good solution to the dereverberation will benefit many speech signal processing technologies.

In this paper, Sect. 2 introduces the mathematical model of the reverberation signal; Sect. 3 introduces the reverberation cancellation algorithm; Sect. 4 introduces the reverberation suppression algorithm; Sect. 5 introduces other comprehensive algorithms; Sect. 6 elaborates the future research directions and give a conclusion.

2 Mathematical Model of Reverberation Signal

The reverberant speech is calculated by convolution of the clean speech and a room impulse response (RIR):

$$y(n) = h(n)^{*}x(n) \tag{1}$$

X. Shao et al. (Eds.): CSMT 2020, LNEE 761, pp. 134–146, 2021.
https://doi.org/10.1007/978-981-16-1649-5_12

where $y(n)$ is the reverberant speech, $h(n)$ is the RIR function, and $x(n)$ is the dean speech. Eliminating the influence of convolution is the main task of speech dereverberation.

According to whether RIR needs to be estimated, algorithms using the statistical acoustic model can be divided into two categories: reverberation cancellation and reverberation suppression, they are discussed separately in the next two chapters.

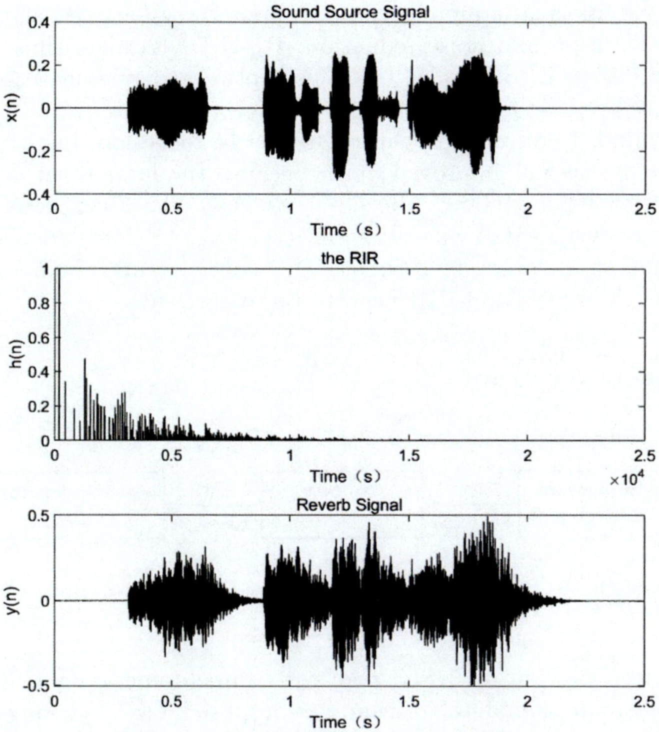

Fig. 1. A reverb signal is convoluted by clean speech signal and the RIR

3 Reverberation Cancellation

The famous multiple-input/output inverse-filtering theorem (MINT) method proposed in 1988 [3] proved the feasibility of reverberation cancellation. Under the premise of known RIR, deconvolution could be performed through an inverse filter, and the clean speech signal can be restored without distortion. However, since RIR is very sensitive to the environment, accurate RIR cannot be measured in real time as a known condition in practical. To solve this problem, blind deconvolution methods and complex cepstrum filters are used.

3.1 Blind Deconvolution

In blind deconvolution algorithms, some information is used to estimate RIR and achieve dereverberation using an adaptive inverse filter. A classical method is using the correlation matrix between multi-channel microphone signals[4]. Based on this, preference [5] studied the process of introducing deviations when using different window functions to truncate the filter impulse response. With the cost of 63% increased computational complexity, the signal-to-noise ratio (SNR) can be increased from −0.6 dB to 9.9 dB after the process. Preference [6] taking the noise statistical information into account, proposed regularized partial MINT for joint dereverberation and noise reduction (RPM-DNR) algorithm, a weighting parameter is designed to balance the effect between dereverberation and noise reduction.

Using a blind deconvolution can reduce early reflection, but the subjective sense of hearing has not improved much because the human ear cannot detect the sound delay below 100ms. More studies make the blind deconvolution as one step in dereverberation, such as the two-stage dereverberation algorithm introduced below. In order to estimate RIR more accurately, more statistical characteristics of reverb and RIR need to be researched.

3.2 Complex Cepstrum

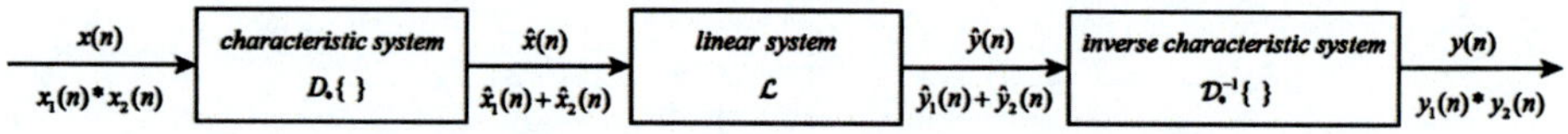

Fig. 2. Canonic Form for Homomorphic Convolution

The blind deconvolution discussed above is a time-domain algorithm. In 1975, a complex cepstrum domain algorithm was proposed [7]. The complex cepstrum of the signal $x(n)$ is expressed as:

$$\hat{x}(t) = \mathcal{F}^{-1}\{\log \mathcal{F}[x(t)]\} \tag{2}$$

Complex cepstrum is a convolutional homomorphic system that can convert convolution in the time domain into addition operation in complex cepstrum domain. The conversion to the complex cepstrum of Eq. (1) can be described as:

$$\hat{y}(n) = \hat{h}(n) + \hat{x}(n) \tag{3}$$

The complex cepstrum of reverberate signal $\hat{y}(n)$ is a bounded, infinite attenuate sequence. The nonlinear process of log operation causes $|\hat{y}(n)|$ to attenuate rapidly with the increase of n. As the distribution of complex cepstrum $\hat{x}(n)$ of the clean speech is dose to the zero points, while the complex cepstrum $\hat{h}(n)$ of the RIR is mainly distributed away from the zero points, the effect of $\hat{h}(n)$ can be eliminated by a low-pass filter in the complex cepstrum domain. By converting the deconvolution operation into subtraction, the calculation complexity can be greatly reduced.

The complex cepstrum algorithm is suitable for clean speech and RIR are far away in the complex cepstrum domain. However, as a nonlinear transformation, converting the signal to the complex cepstrum domain filtering will cause some frequency distortion.

In preference [8], researchers use the mean subtraction of complex cepstrum and then discuss the effect of different window functions on the speech frame. Although it is simple to implement, it also has a phase ambiguity problem, which directly points out two difficulties that the traditional complex cepstrum domain dereverberation have:

1. It is difficult to find the best window function type, and the parameters of complex cepstrum domain filters are hard to determine;
2. The RIR is usually not the minimum phase in practical [9], which will cause phase ambiguity in the complex cepstrum. In this condition, the log phase of the two complex numbers' product does not satisfy the additive property, so it is not easy to reconstruct the original signal.

To solve the first problem, preference [10] studied and determined the parameters such as the maximum cut off point of the low pass filter in the complex cepstrum domain, the transition bandwidth, and the curve characteristics of the transition band. The maximum cut off point of the low pass filter is irrelevant to the reverberation time. Adding a Gaussian window before complex cepstrum domain filtering can improve the dereverberation effect. For the second problem, the researchers used the minimum phase decomposition method to optimize. A causal minimum phase LTI system with a rational transfer function $H(z)$ is stable, which means all poles of $H(z)$ are inside the unit circle of the z-plane. To be implemented in a practical situation, we need to convert the system to a minimum phase with a stable solution in complex cepstrum. Any rational function system can be expressed as a combination of a minimum phase part $H_{(z)}$ and an all-pass part $H_{ap}(z)$ [11]:

$$H(z) = H_{\min}(z)H_{ap}(z) \tag{4}$$

By decomposing the signal into these two components, preference [12] applied a complex cepstrum filter to the minimum phase component, as well as reconstructed the signal with an all-pass component, which solved the two major problems mentioned above. The all-pass component appears as the first prominent positive peak in the complex cepstrum domain, which can accurately maintain the phase information. That is, the phase information will not be lost but can be stored in the all-pass component [13]. In fact, the spectral phase's processing has been largely neglected in speech signal processing during the past decades until recently, more studies have found that there is actually much useful information in phase, and the use of this information can promote many fields of speech processing [14,15]. Therefore, the work direction for the future should increase the use of phase information.

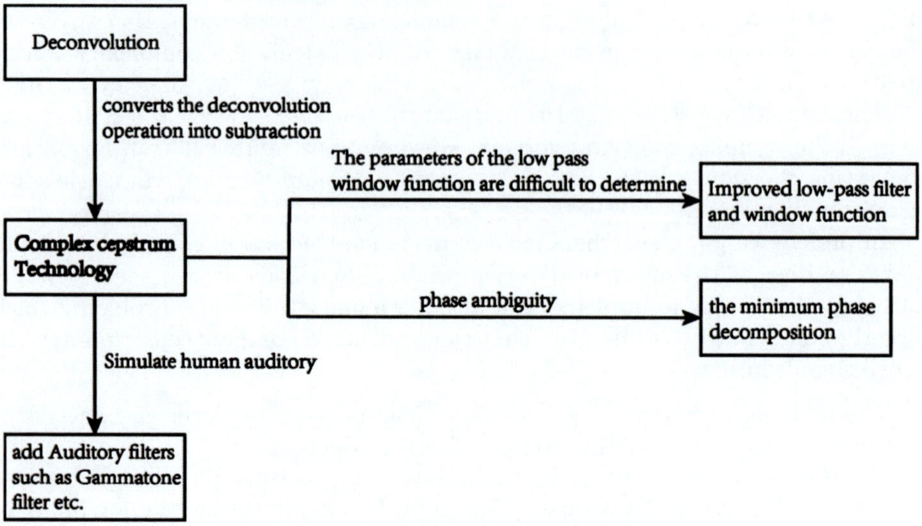

Fig. 3. The development of dereverberation algorithm using complex cepstrum

4 Reverberation Suppression

These algorithms using speech characteristics to analyze the effect of reverb, process the speech, suppressing the reverb effect by relevant features of RIR. Although suppression cannot eliminate the reverb completely, it still can improve the SNR. Reverberation suppression is more widely used than reverberation cancellation in practical because it is easier to implement for avoiding estimating the time-varying RIR. Algorithms based on the linear prediction (LP) residual and spectral subtraction processing belong to this category.

4.1 Linear Prediction Residual

LP residual signal processing is a speech enhancement technology that can effectively remove additive noise. The premise of applying LP to speech dereverberation is to assume that the reverb is mainly limited to prediction residuals and has little effect on coefficients of linear predictive coding (LPC). The late reflections are relatively white in the LP residual domain [16]. Therefore, we regard the clean speech and the early reflections as expected signal and estimate them using linear prediction to suppress the late reverberation. The late reflections in the reverberation signal can be expressed separately as:

$$y^m(n) = d^m(n) + \sum_{p=0}^{P} c_p^m x^m(n - D - p) \tag{5}$$

Where m is the number of microphones, $d^m(n)$ is the direct sound with early reflections, D is the sampling point to distinguish the early reflections and the late reflections. By finding a set of coefficients $c^m(n)$, the desired signal can be recovered. Using weighted Linear Prediction Error (WPE) is also one of the most mainstream dereverberation algorithms in practical applications. WPE performs long-term linear prediction at each frequency point of the short time Fourier transform (STFT). Assuming that late reverberation is not related to clean speech with early reflections, then

$$\hat{s}_n[f] = y_n[f] - \sum_{\tau=D}^{D+O} G_\tau[f]^H y_{n+\tau}[f] \tag{6}$$

Where n is the serial number of time frame, $y_n[f]$ are the coefficient vectors of the STFT at different frequency f, S_n is the prediction error vector, G_τ is the complex-valued square matrix of the prediction matrix, symbol $(\cdot)^H$ represents the conjugate transpose, and D is the length of prediction step, O is the order of the prediction filter. The prediction step D is usually set to 2 or 3 instead of 1 in the WPE algorithm to reduce the excessive decorrelation effect of LP, but a certain coloration will be introduced accordingly [17]. The classic WPE algorithm's performance largely depends on the estimation accuracy of the expected signal power spectral density (PSD). When the observed signal lasts long, every time the iteration in the WPE adaptive algorithm can improve the PSD estimation, otherwise, the PSD estimation will deviate largely, causing the dereverberation performance to decrease accordingly. To improve this problem, preference [18] incorporated a deep neural network (DNN) based spectrum estimator into the WPE framework so that PSD can be reliably estimated from very short observation signals. Experiments showed the processed speech had improved ASR performance compared to the traditional WPE method. Based on this, preference [19] approximated the inter frame correlation (IFC) of STFT and used it to derive WPE, the WER can go down by about 1% compared with the traditional WPE method. Preference [20] unified the WPE method and a variant of the minimum variance distortionless response (MVDR) beamformer into a single convolutional beamformer, reducing the WER by 3.84% on average than the WPE method. The revolution in machine learning technology has made remarkable achievements in various fields related to speech processing [21].

Preference [22] optimized the convex function to improve far-field speech recognition, and a six-microphone array is used based on the traditional multi-channel LP algorithm. The result showed that when the receiving point is 5m away from the sound source, the WER is about 12%, about 2% lower than before. Preference [23] used a Mixed Autoregressive (MAR) reverberation model based on LP, in which a time-varying first-order Markov model is used to estimate its coefficients, combined with Kalman filtering for noise reduction. The number of microphones can be controlled within a certain range to achieve real-time processing requirements. After processing, the speech SNR is improved by about 0.5 dB. Preference [24] assumed that both the prediction coefficient and the residual signal are sparse, and the processed signal is relatively improved.

Since auto regressive (AR) models are often used, to ensure the accuracy of linear prediction, the AR model's order generally needs to be greater than 9. To optimize and simplify the algorithm, finding a balance between system performance and calculation complexity is an important research direction of LP residual algorithms.

4.2 Spectral Subtraction

Preference [25] proposed that spectral subtraction can be used to achieve speech dereverberation. This method estimates the energy spectrum of reverberation, then converts the signal into the energy spectrum to subtract it. Spectral subtraction treats late reflection reverberation as additive noise:

$$y(n) = x(n) + d(n) \tag{7}$$

Like the LP residual algorithm, this method sees the sum of clean speech and early reflections as the desired signal and mainly removes late reflections. Assuming that $x(n)$ and $d(n)$ are independent of each other, after performing the STFT on Eq. (7), the spectral subtraction can be described as:

$$|\hat{X}(\omega)| = (|\hat{Y}(\omega)| - |\hat{D}(\omega)|)e^{j\phi_y(\omega)} \tag{8}$$

By estimated the power spectrum of late reflections and subtract it, the effects of suppressing reverberation can be achieved. Because reverberation has different effects on different frequency bands, and the algorithm estimates the noise inaccurately in low SNR, the residual noise will fluctuate in a narrow band producing nonlinear distortion. This noise will interfere with the desired signal and affect the quality of speech. The improved algorithm used statistical acoustics' characteristics to find the expected minimum value of certain distortion metrics between the clean speech and the estimated signal. However, there is no unified voice statistical model or unified distortion in such algorithms measure for now. The Polack reverberation statistical model is used in preference [26]; The super-Gaussian prior speech model and the Laplacian noise model is used in preference [27]. Some studies used the minimum mean square error (MSE) as the distortion metric, while some studies used the estimated clean speech phase and multi-band spectral subtraction to estimate the amplitude to solve the "music noise" problem [28].

Spectral subtraction is not sensitive to RIR's fluctuation, and it is not suitable for eliminating early reflections. Excessive reduction of the energy spectrum will cause nonlinear distortion, known as "music noise", which will decrease the speech quality. Therefore, spectral subtraction is not suitable for standalone. It can be improved when combined with other algorithms. For example, in the two-stage dereverberation algorithm (Sect. 5.1), it is used in the second stage of processing.

4.3 Beamforming

Beamforming is one of the main research directions in array signal processing. It is often used to extract specific sound sources in noisy environments, filter out noise and sound from unexpected directions. With the development of 5G technology, massive MIMO systems such as microphone array have become a hot research topic. Since RIR is related to the position, and the microphone array system has multiple microphones, it can record the position information in the space and has more available matrix information than the single-channel system. Therefore, the dereverberation algorithm using the microphone array can usually achieve better results than single-channel systems.

The signals received by the microphone array are decomposed into the minimum phase and the all-pass component, filtering the minimum phase component on the cepstrum domain to eliminate reverberation, and then recombine multiple processed components to reconstruct speech. This algorithm's effect is improved compared with complex cepstrum algorithms, and it is very suitable for the time-varying RIR; but when the reflection coefficient increases above a certain threshold, the beamforming array's amplitude will be reduced [13]. Preference [29] proposed a multi-channel inverse filtering for RIR based on skewness, which does not need prior knowledge of RIR or direction of arrival (DOA). This method used a non-Gaussian maximum criterion to implement blind inverse filtering and had a better effect in a strong reverb situation. Preference [20] used the MVDR beamforming and multi-channel LP, unified into a convolutional beamformer to achieve the best integration of noise reduction and dereverberation. This method greatly improved speech enhancement performance and reduced the speech recognition WER, but parameters such as the target signal's direction angle need to be estimated. Generally, the speech's main information varies sparsely distributed on the frequency spectrum and contains only a limited number of harmonics. If using a beamforming filter such as DOA, MVDR, it would cover a wide frequency band, generate unwilling noise in the signal band out of interest. Some filters based on harmonic models have been proposed and applied to dereverberation [30–32], but these harmonic models have not been widely used in beamforming technology, which can be further studied in the future. The naturalness of the beamforming method's subjective evaluation is the best, but the dereverberation effect is also the worst among all objective evaluation indicators. This showed that delay-weighted summation and other beamforming methods mainly eliminate the effect of additive noise; eliminating convolution noise such as reverberation is not obvious. It is more suitable for noise reduction preprocessing of other dereverberation algorithms.

5 Other Comprehensive Algorithms

This chapter introduces some novel composite algorithms with good dereverberation effects.

5.1 Classic Two-Stage Dereverberation

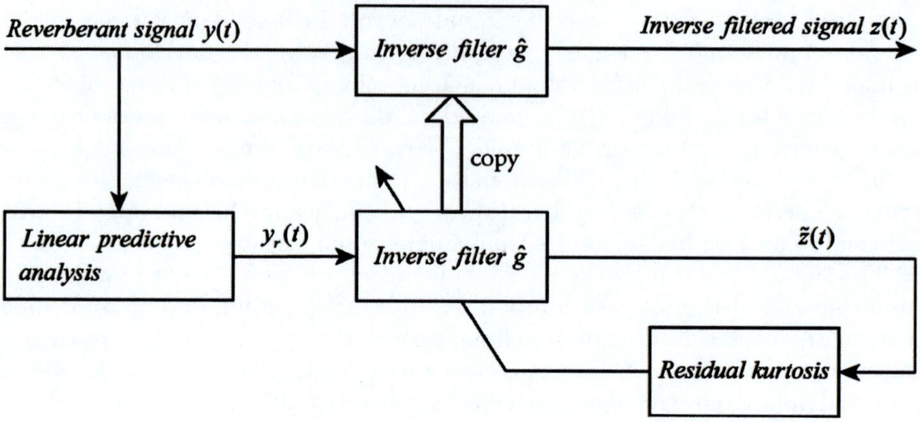

Fig. 4. Process structure of the first stage in the two-stage algorithm

The quality of reverb speech depends on two physical variables: SRR and reverberation time T_{60}. Preference [11] proposed a two-stage dereverberation method to deal with these two quantities separately: In the first stage, an inverse filter is estimated to increase the SRR, the second stage used spectral subtraction to suppress the tail of reverberation. In the first stage, the RIR inverse filter of fixed length is estimated by maximizing the kurtosis of the LP residual, as shown in Fig. 4, for the kurtosis of the LP residual of clean speech is the highest. After processing, the SRR is rose from −9.8 dB to 2.4 dB. This stage's effect is similar to the effect of moving the sound source closer to the microphone. In the second stage, the equalized RIR is divided into early and late parts, then uses the spectral subtraction process to target the late reflections. Speech after two-stage processing, the average SNR can gain by 4.82 dB. The inverse filtering of this algorithm is only effective when T_{60} is between 2–4 s. If the reverberation time is longer than 4s, the target function of adaptive inverse filtering based on kurtosis will have many saddle points, resulting in inaccurate estimation. Preference[33] analyzed the LP residual characteristics of reverberant speech, proved that it is effective to use the third-order moment in the statistical feature to suppress reverb. Furthermore, through comparative experiments, it is found that the skewness maximization based on the LP residuals is better and more robust than the kurtosis maximization. The two-stage algorithm based on maximizing LP's residual skewness has a significant improvement, especially with long reverberation time.

The two-stage dereverberation algorithm has been combined with deep neural networks (DNN) to get a better effect in recent years. Express the spectrum enhancement or separation problem as a supervised learning problem and then use DNN for supervised learning. Preference [34] combined the DNN framework and proposed using the Hierarchical structure of the extreme learning machine (HELM) learning model for speech dereverberation. This model does not adjust the feature extraction layer parameters but only estimated the conversion matrix based on the training data, which is very suitable for embedded applications and mobile devices; instead of the inverse filter estimation, preference [35] used the classical dictionary training K-SVD algorithm in machine learning; preference [36] used a DNN with 3 hidden layers to estimate the ideal ratio mask (IRM) corresponding to the first stage in the classic two-stage algorithm. In the second stage, a MSE normalization is used instead of the percentage normalization, which could preserve more spectral details, and it is more conducive to restore clean speech. The experiment showed that the algorithm using DNN combined with a two-stage algorithm has an average PESQ score of 0.07 points higher than before. When $T_{60} = 0.3$ s, the average direct reverberation ratio (DRR) can reach 4.96 dB.

5.2 CDR Estimation

In 2011, Jeub et al. proposed a novel algorithm for dereverberation using two omnidirectional microphones [33]. The author defined a coherent-to-diffuse power ratio (CDR) equation:

$$\Psi\left(e^{j\Omega}\right) = \frac{\Phi_c\left(e^{j\Omega}\right)}{\Phi_d\left(e^{j\Omega}\right)} \tag{9}$$

The coherent and diffuse power spectrum between two received signals is $\Phi_c\left(e^{j\Omega}\right)$ and $\Phi_d\left(e^{j\Omega}\right)$ respectively. For the same signal received in different positions in a uniformly diffused sound field, we can assume that the direct signals are coherent, and the reverb signals are incoherent by the two microphones. The CDR function is used to describe the correlation ratio between the clean and reverberant signals, then remove reverb by estimating the correlation according to Eq. (9). Preference [37] proposed three new CDR estimation methods, summarized and compared all seven kinds of CDR estimation methods, including proposed before, four of which are DOA-dependent CDR estimation methods. The CDR estimation method proposed by Jeub belongs to this category. There are two other DOA-independent CDR estimators and one estimator that do not need the noise coherence information. There is an infinite number of unbiased estimates for CDR DOA-independent estimators, but only one DOA-dependent unbiased CDR estimator can be determined. Experimental results showed that using this equation:

$$\widehat{CDR}(l,f) = \frac{1 - \tilde{\Gamma}_n \cos\left(\arg\left(\tilde{\Gamma}_s\right)\right)}{\left|\tilde{\Gamma}_n - \tilde{\Gamma}_s\right|} \left|\frac{\tilde{\Gamma}_s^*\left(\tilde{\Gamma}_n - \tilde{\Gamma}_x\right)}{\mathrm{Re}\left\{\tilde{\Gamma}_s^*\tilde{\Gamma}_x\right\} - 1}\right| \tag{10}$$

to estimate the CDR, both the speech recognition WER and the PESQ score are the best: the speech recognition rate reached 90.0%,and the PESQ score reached 1.76 compared with Thiergart's estimator:

$$\widehat{CDR}_{\text{Thiergart}}(l,f) = \mathrm{Re}\left\{\frac{\tilde{\Gamma}_n - \tilde{\Gamma}_x}{\tilde{\Gamma}_x - \tilde{\Gamma}_s}\right\} \tag{11}$$

This estimator scored the lowest in the same room test: the speech recognition rate was 86.2%, and the PESQ score was 1.46. In the Eqs. (10) and (11), $\tilde{\Gamma}_n$ is the estimation of the coherent function of the noise signal in the two microphones, $\tilde{\Gamma}_s$ is the estimation of the coherent function of the dean speech signal, and $\tilde{\Gamma}_x$ is the estimation of the coherent function of the received signal. The CDR dereverberation algorithm is novel and effective, the use of two microphones compensates for the shortcomings of lacking spatial information in a single microphone system. It is also relatively simplified compared to the microphone array system, so it is easier to implement. Nowadays, most mobile phones have two microphones, so this algorithm is likely to be widely applied to many mobile devices in the future. However, the currently proposed CDR estimation models are not robust enough in practice, and part of them are biased estimates. This algorithm's future research direction can further optimize the CDR function estimation model, effectively correct the biased estimation and improve the robustness.

6 Conclusion

An overview of speech dereverberation is given in this paper. According to whether RIR needs to be estimated, speech dereverberation algorithms can be divided into reverberation cancellation and reverberation suppression. The reverberation cancellation uses the inverse filter for deconvolution operation, is limited by the complexity of calculation and sensitivity to noise; the actual application is not as wide as the reverberation suppression. Older dereverberation algorithms are often given a priori assumptions to simplify the statistical models, resulting in low confidence. With the rapid development of computing ability, the decreasing price of storage devices and the continuous emergence of big data, many algorithms that were considered too complex to be calculated can be easily implemented. Therefore, algorithms that are sensitive to noise and RIR changes, such as reverberation cancellation, can consider more time-varying information to design more accurate and complex description models. Even if the calculation complexity increases, it is possible to achieve real-time calculation as the device's computing performance increases.

On the other hand, wearable devices will be used in an outdoor environment with harsh acoustic environments and unstable sound fields for a long time, the dereverberation algorithm with high robustness and high noise reduction is one of the key research directions.

From the perspective of the dereverberation effect, the microphone array system can make full use of spatial information, which is still an important area of future research. The combination of adaptive technology and other technologies is the main research point in the future. The function model involved in the detailed optimization algorithm, including auditory filter bank, harmonic model, CDR estimation, etc. need further optimization, and a good phase theory needs to be studied deeply.

References

1. Huang, Y., Benesty, J., Chen, J.: A blind channel identification-based two-stage approach to separation and dereverberation of speech signals in a reverberant environment. IEEE Trans. Speech Audio Process. **13**(5), 882–895 (2005)
2. Kinoshita, K., et al.: A summary of the reverb challenge: state-of-the-art and remaining challenges in reverberant speech processing research. EURASIP J. Adv. Signal Process. **2016**(1), 7 (2016)
3. Miyoshi, M., Kaneda, Y.: Inverse filtering of room acoustics. IEEE Trans. Acoustics Speech Signal Process. **36**(2), 145–152 (1988)
4. Furuya, K.: Noise reduction and dereverberation using correlation matrix based on the multiple-input/output inverse-filtering theorem (MINT). In: International Workshop on Hands-Free Speech Communication, pp. 59–62 (2001)
5. Douglas, S.C., Sawada, H., Makino, S.: Natural gradient multichannel blind deconvolution and speech separation using causal FIR filters. IEEE Trans. Speech Audio Process. **13**(1), 92–104 (2004)
6. Kodrasi, I., Doclo, S.: Joint dereverberation and noise reduction based on acoustic multi-channel equalization. IEEE-ACM Trans. Audio Speech Lang. **24**(4), 680–693 (2016)
7. Oppenheim, A.V., Schafer, R.W.: Digital Signal Processing. Prentice-Hall, Inc., Upper Saddle River (1975)
8. Bees, D., Blostein, M., Kabal, P.: Reverberant speech enhancement using cepstral processing. In: IEEE International Conference on Acoustics, Speech, and Signal Processing, pp. 977–980. IEEE Computer Society (1991)
9. Neely, S.T., Allen, J.B.: Invertibility of a room impulse response. J. Acoust. Soc. Am. **66**(1), 165–169 (1979)
10. Zhang, D.h., Chen, G.y.: Speech signal dereverberation with cepstral processing. Tech. Acoust. (1), 12 (2009)
11. Wu, M., Wang, D.: A two-stage algorithm for one-microphone reverberant speech enhancement. IEEE Trans. Audio Speech Lang. Process. **14**(3), 774–784 (2006)
12. Qi-penga, L., Ronga, K., Yuan-yuana, S., Ji-huaa, G., He-mingb, Z., Zhia, T.: Dereverberation based on minimum phase decomposition. Commun. Technol. **6**, 7–13 (2011)
13. Liu, Q.G., Champagne, B., Kabal, P.: A microphone array processing technique for speech enhancement in a reverberant space. Speech Commun. **18**(4), 317–334 (1996)
14. Mowlaee, P., Saeidi, R., Stylianou, Y.: Advances in phase-aware signal processing in speech communication. Speech Commun. **81**, 1–29 (2016)
15. Paliwal, K., Wójcicki, K., Shannon, B.: The importance of phase in speech enhancement. Speech Commun. **53**(4), 465–494 (2011)
16. Peng, R., Tan, Z.H., Li, X., Zheng, C.: A perceptually motivated lp residual estimator in noisy and reverberant environments. Speech Commun. **96**, 129–141 (2018)
17. Yoshioka, T., Nakatani, T., Miyoshi, M., Okuno, H.G.: Blind separation and dereverberation of speech mixtures by joint optimization. IEEE Trans. Audio Speech Lang. Process. **19**(1), 69–84 (2010)
18. Kinoshita, K., Delcroix, M., Kwon, H., Mori, T., Nakatani, T.: Neural network-based spectrum estimation for online WPE dereverberation. In: Interspeech, pp. 384–388 (2017)
19. Parchami, M., Zhu, W.P., Champagne, B.: Speech dereverberation using weighted prediction error with correlated inter-frame speech components. Speech Commun. **87**, 49–57 (2017)

20. Nakatani, T., Kinoshita, K.: A unified convolutional beamformer for simultaneous denoising and dereverberation. IEEE Signal Process. Lett. **26**(6), 903–907 (2019)
21. Zhang, X., Li, Y., Zheng, Cao, T., Sun, M., Min, G.: Research progress and prospect of speech dereverberation technology. J. Acquisit. Process. Data **32**(6), 1069–1081 (2017)
22. Giacobello, D., Jensen, T.L.: Speech dereverberation based on convex optimization algorithms for group sparse linear prediction. In: 2018 IEEE International Conference on Acoustics, Speech and Signal Processing (ICASSP), pp. 446–450. IEEE (2018)
23. Braun, S., Habets, E.A.: Linear prediction-based online dereverberation and noise reduction using alternating Kalman filters. IEEE-ACM Trans. Audio Speech Lang. **26**(6), 1119–1129 (2018)
24. Mousavi, L., Razzazi, F., Haghbin, A.: Blind speech dereverberation using sparse decomposition and multi-channel linear prediction. Int. J. Speech Technol. **22**(3), 729–738 (2019)
25. Lebart, K., Boucher, J.M., Denbigh, P.N.: A new method based on spectral subtraction for speech dereverberation. Acta Acust. United Acust. **87**(3), 359–366 (2001)
26. Giesbrecht, D., Hetherington, P.: Reverberation estimation and suppression system (2012). US Patent 8,284,947
27. Martin, R.: Speech enhancement based on minimum mean-square error estimation and supergaussian priors. IEEE Trans. Speech Audio Process. **13**(5), 845–856 (2005)
28. Zhen, L., Wenjin, W., Qin, Z., Hui, R.: Multi-band spectral subtraction of speech enhancement based on maximum posteriori phase estimation. J. Electron. Inf. Technol. **39**(9), 2282–2286 (2017)
29. Guo, Y., Peng, R., Zheng, C., Li, X.: Maximum skewness-based multichannel inverse filtering for speech dereverberation. J. Appl. Acoust. **38**(1), 58–67 (2019)
30. Christensen, M.G., Jakobsson, A.: Multi-pitch estimation. Synth. Lect. Speech Audio Process. **5**(1), 1–160 (2009)
31. Harvey, B., O'Young, S.: A harmonic spectral beamformer for the enhanced localization of propeller-driven aircraft. J. Unmanned Veh. Syst. **7**(2), 156–174 (2019)
32. Schmidt, A., Löllmann, H.W., Kellermann, W.: A novel ego-noise suppression algorithm for acoustic signal enhancement in autonomous systems. In: 2018 IEEE International Conference on Acoustics, Speech and Signal Processing (ICASSP), pp. 6583–6587. IEEE (2018)
33. Mosayyebpour, S., Sheikhzadeh, H., Gulliver, T.A., Esmaeili, M.: Single-microphone LP residual skewness-based inverse filtering of the room impulse response. IEEE Trans. Audio Speech Lang. Process. **20**(5), 1617–1632 (2012)
34. Hussain, T., Siniscalchi, S.M., Wang, H.L.S., Tsao, Y., Mario, S.V., Liao, W.H.: Ensemble hierarchical extreme learning machine for speech dereverberation. IEEE Trans. Cogn. Dev. Syst. 1–15 (2019)
35. Kilis, N., Mitianoudis, N.: A novel scheme for single-channel speech dereverberation. In: Acoustics, vol. 1, pp. 711–725. Multidisciplinary Digital Publishing Institute (2019)
36. Zhao, Y., Wang, Z.Q., Wang, D.: Two-stage deep learning for noisy-reverberant speech enhancement. IEEE-ACM Trans. Audio Speech Lang. **27**(1), 53–62 (2018)
37. Schwarz, A., Kellermann, W.: Coherent-to-diffuse power ratio estimation for dereverberation. IEEE-ACM Trans. Audio Speech Lang. **23**(6), 1006–1018 (2015)

Animal Sound Analysis

Detection of Basic Emotions from Cats' Meowing

Qianlong Shou, Yumeng Xu, Junjun Jiang, Min Huang(✉), and Zhongzhe Xiao(✉)

School of Optoelectronic Science and Engineering, Soochow University, Suzhou 215006, Jiangsu, China
{hmin,xiaozhongzhe}@suda.edu.cn

Abstract. Basic emotional states in valence sense as positive, neutral, and negative are studied with automatic classification on cats' meowing signals, aiming to help human-cat interaction and human emotion regulation by pets keeping. The ground truth of meowing samples is marked by subjective evaluation from multiple raters with the help of cats' facial expression, body movement, and interaction with cat owners in video clips. Acoustic features extracted from voice energy, zero crossing rate, and MFCC are proved to be effective in cats' emotion recognition. The highest accuracy reaches 97.40% on selected best feature subset with LogitBoost model.

Keywords: Cats emotions · Acoustic features · Recognition

1 Introduction

Voice, as an effective communication style, plays an essential role in the expression of feelings. In recent years, researchers have yielded numerous remarkable results in the emotional analysis of speech for humans [1–3] and various speech emotion datasets [4] have been obtained. Furthermore, the recognition for emotions of human has achieved a much higher accuracy. Overall, great progress has been made in the study of emotions in human voice. However, few studies have focused on the analysis of emotions for animals, and there are a few affective computing techniques to recognize the emotions for mammals except humans.

Some researchers have studied the barking of dogs and analyzed the emotions contained in dog barkings. Acoustic characteristics have been discussed for the recognition of dogs by their barkings [5] and many features have been proposed, as well as methods, while emotions of cats, who are also important accompany pets of humans, are not yet studied thoroughly with automatic analysis. A good model in recognition of cats' emotions, will greatly help human, especially new owners of cats, to quickly develop a better interaction with their pets, and make the most advantage of cat keeping, for accompany, or even emotion regulation for human (cat owner).

X. Shao et al. (Eds.): CSMT 2020, LNEE 761, pp. 149–160, 2021.
https://doi.org/10.1007/978-981-16-1649-5_13

From the related studies on human emotions expressed by voice, one existing problem is that the emotion categories never reached any universal agreement. Relatively commonly accepted emotion taxonomies include Ekman's "big six" [6], or two-dimensional model with valence and arousal [7]. Application dependent definition of emotion categories is also a common manner, such as in the case of several widely used emotional speech datasets [8–10]. Although there are currently very few studies on cats' emotions, similar investigation has been made with dogs as behaviour and emotion models of companion robots [5,11–13]. For example, application dependent emotions as happiness, despair, fear, anger, and surprise are used in [12]. In this work on cats' emotions by meowing voice, we choose to use a simple way as the starter, with three states in valence sense as positive, neutral, and negative, to describe the cats' most basic emotions.

In the machine learning based approaches, a dataset with reliable labeling of ground truth to each sample is the essential basis. For example, in the work with dog barkings, perception tests indicate that acoustic parameters, including tonality, pitch and inter-bark time intervals, are strongly related with emotions and affect greatly on listeners' judgment [14–17]. In building meowing dataset in this work, subjective evaluation with human judgement will also be used, while with the help of video contents including cat facial expressions, body movement, *etc.*., because human judgement of cats' emotions only by meowing voice is not a practical activity for most persons.

The rest of the paper is organized as follows. Section 2 introduces the construction process of the dataset of cats' meowing. Section 3 describes the process of emotional feature extraction and conducts the feature dimensionality reduction to avoid curse of dimensionality. Section 4 gives a set of experiments for evaluation. Finally, Sect. 5 concludes the paper and presents the future work.

2 Dataset Construction of Cats' Meowing

We aim to perform an automatic detection of cat's basic emotions from cats' meowing in a data driven manner. Thus, to collect a dataset with sufficient cats' meowing samples with reliable labelling is an essential preparation. The collected meowing samples can then be regarded as cats' "language" in the cat emotion detection.

There are two basic concerns in the construction of this dataset. First, meowing samples from only one or two cats will introduce great influence of the cat individual, and the common clues in cats' emotions expressing by voice cannot be fully discovered. Second, with only the audio signal of cats' meowing, we cannot accurately judge the cats' emotional states. The cats' facial expressions, body movements, and the surrounding situations including their interactions with their owners will help greatly for the judgement. For the above two reasons, video clips from public websites/apps are chosen as the resource of cats' meowing samples in this work. The resource websites we used in data collection include Iqiyi, Tencent Video, Bilibili, Haokan Video, Wesee Video, etc. A lot of "cat persons" are sharing their daily interactions with their cats, with labels or even detailed explanations to the shared scenes. These sharings facilitate persons who like "cyber cat petting" to satisfy their cataholic. By collecting samples from these video websites, we can obtain cats' scenes from a large number of different cats. The labels and explanations also make our data collection much easier, because the owners of the cats know their cats very well and the labels can be seen to be reliable reflecting the cats' emotional states. The construction process of the dataset is illustrated in Fig. 1, from the video clips with owners' labels and explanations.

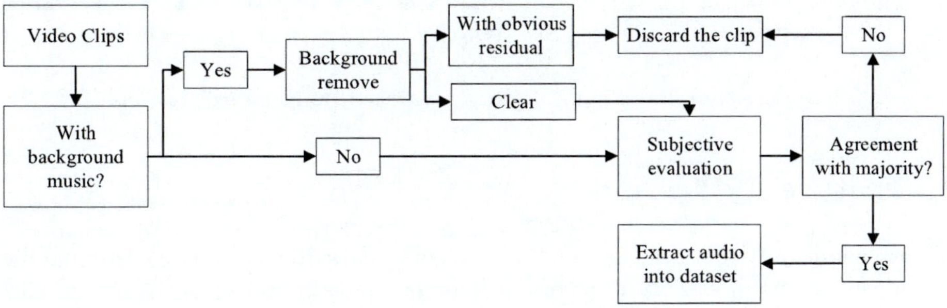

Fig. 1. Flow chart of constructing cats' meowing dataset

One of the problems in the meowing samples collection is that cat owners usually add background music with their uploaded videos, while the background music will significantly influence the analysis of cats' voice. In this case, we first make a preprocessing with Adobe Audition to remove it. If the obvious residual of background can still be heard after the removal, the video clip will be discarded. Video clips with clear removal of background music, together with clips without any background music, are sent to subjective evaluation in the next step.

In subjective evaluation, several raters were asked to evaluate the emotional states of cats. To make the evaluation not too dispersed, we only set three categories of emotions in valence sense as positive, neutral, and negative. Positive emotions include happy, contentment when they get food or play with their owners, sometimes the cat will make snoring like sound to express their satisfaction. Negative emotions include the states such as hunger, scared, anger, etc. The usual states are regarded as neutral. The raters make the evaluations by watching videos, including the cats' facial expressions, body movements, and interactions with cat owners as their basis of judgement, and if the cat owners provided labels or explanation, this will also be very important evidence for the raters. Examples of cats' facial expressions are shown in Fig. 2. Not all raters have to evaluate all collected video clips, but we ensured that each clip received evaluations from at least 3 raters. When majority of raters give consistent judgement, this clip will be marked with the corresponding positive, neutral, or negative label as the ground truth, and the audio part is extracted into the dataset. If no majority judgement exists, the clip will be discarded.

Totally 566 samples are kept in our collected dataset, with 179 positive samples, 141 neutral samples, and 246 negative samples.

(a) Positive (b) Negative (c) Neutral

Fig. 2. Examples of cat emotions presented by facial expressions

3 Feature Extraction

Effective features that can express the characteristics of cat emotions from meowing voice are one of essential factors in machine learning based recognition. Currently, there are very few studies on feature analysis of cat meowing for emotions, we proposed in this paper to adopt experience from emotion recognition works on speech, music, or other common audio signals. A good choice is to use the feature set provided by the challenges of INTERSPEECH, such as the emotion challenge in 2009 [18], or more comprehensive paralinguistics challenges in 2010 and 2013 [19,20]. These feature sets have been proved in a number of work concerning human speech emotion [21], and could be a good starter for this investigation of cat meowing emotion.

Concerning that we only collected several hundreds samples of emotional cats' meowing, which only form into a small scaled dataset, high dimensional feature sets will cause the problem of overfitting. In order to minimize the impact, the feature set from INTERSPEECH 2009 emotion challenge, which is with the fewest dimension of features in this series of challenge feature sets, is adopted in this work. Three categories of features, as prosody features, sound quality features and spectral features, are contained in this feature sets. The overall extraction of these features to apply 12 statistical functions on 16 low-level descriptors (LLDs) and their first order difference, to result into features, as shown in Fig. 3.

The 16 LLDs are zero-crossing-rate (ZCR), root mean square (RMS) energy, fundamental frequency (F0), harmonic-noise ratio (HNR), and first 12 Mel frequency cepstrum coefficients (MFCCs). The functions to be applied on these LLDs range from first order to higher order statistics, including mean, standard

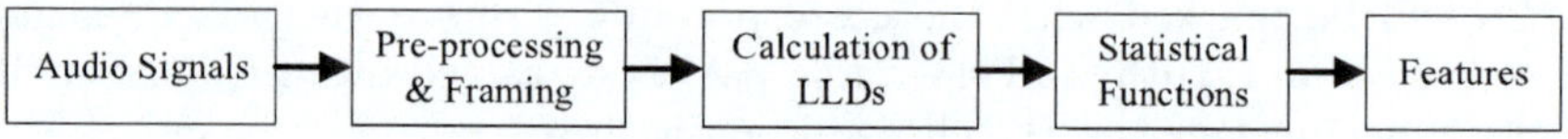

Fig. 3. Feature extraction process

deviation, kurtosis, skewness, maximum and minimum value, relative position, range, and offset and slope of linear regression, together with their mean square error.

The extraction of the above feature set for cats' meowing analysis is based on TUM's open-source openSMILE toolkit [22], with the configuration "emotion_IS09.conf".

4 Automatic Emotion Detection of Meowing

4.1 Experiment Settings

The machine learning approaches for the cats' meowing emotion detection are implemented on WEKA platform [23]. Logistic Regression for classification is chosen as the most basic algorithm in this investigation. As the cats' emotion is not linear, this generalized linear model may not fully present the distinguishing ability of the features, two higher level classifiers based on logistic are used for better performances, as LogitBoost (as in WEKA platform), which uses boosting method based on logistic with maximum likelihood for optimization, and LMT, which builds a tree structure classifier with each node as a logistic model. Beside LMT tree model, another tree model, Random Forest, is also evaluated for comparison.

There is a problem in the collected meowing dataset that the number of samples is extremely unbalanced in each category. This imbalance will significantly influence the reliability of the trained models [24]. Thus, we desampled the negative and positive states, to leave only 141 samples in each category to balance with neutral samples. All the evaluations of models are implemented with 10-fold cross validation, to minimize the bias in dividing such small scaled dataset into training set and test set.

4.2 Classification Results of Cats' Meowing

Automatic classification of cats' emotions from meowing voice is performed with the above mentioned four classifiers, and we present the results in terms of accuracy, AUC (area under ROC curve), and confusion matrices.

The accuracies and kappa statistics from the 4 selected classifiers are compared in Fig. 4. The most basic classifier, logistic, presents relatively poor performance with accuracy of only 63.36%, and AUC as low as 0.56, indicates unreliable emotion detection ability with this method. The compound methods based on logistic, LogitBoost and LMT, get significantly improved accuracies of 94.33% and 86.52%, with AUC of 0.88 and 0.75, respectively. These improvements show that the compound methods fit the cats' emotion detection problem better than the basic logistic methods, and the performance especially benefits from the boosting approach, while the tree structure also helps to get better classification in this task of cats' emotion detection from meowing voice. Another tree based

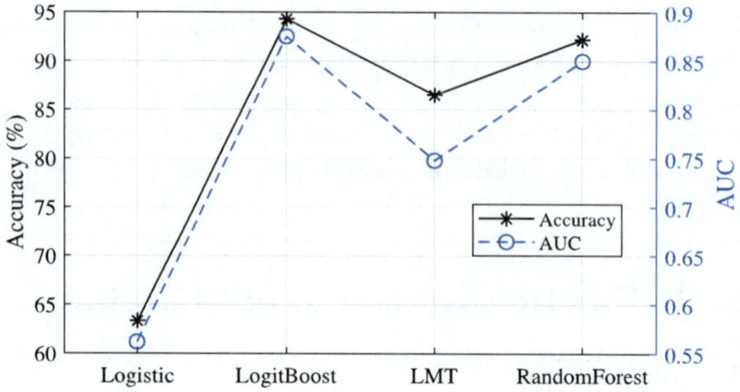

Fig. 4. Accuracies of cats' emotion classification with 4 classifiers

method evaluated here is the Random Forest algorithm, which achieved accuracy of 92.20% with AUC of 0.85. This result is close to that of LogitBoost, and is also highly reliable with high kappa value.

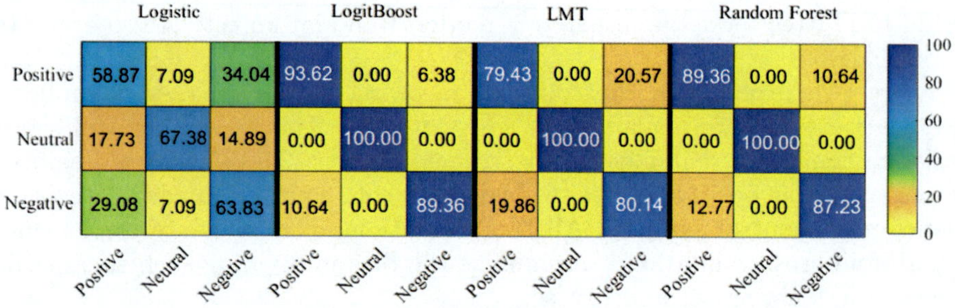

Fig. 5. Confusion matrices of cats' emotion classification (%)

The confusion matrices from the 4 classifiers are shown in Fig. 5, darker colors correspond to higher rates. The worst one, Logistic, presents high confusion to positive or negative from all categories, almost symmetric with positive and negative. A notable phenomenon appears in all other better classifiers that the neutral state is always perfectly classified, and all confusions appear between positive state and negative state. This can be explained by a known fact from human speech emotion that the emotions are easier to be distinguished in arousal dimension than in valence dimension. In this evaluation of cats' emotion from meowing voice, the positive and negative states are defined in valence dimension, while both states present higher arousal than neutral state, thus it leads to the result that the neutral is better recognized than both positive and negative, rather than presented as a middle state between them.

4.3 Further Analysis

From the accuracies of over 90% in cats' emotion classification from LogitBoost and RandomForest, we assume that the cats' emotional states can be detected by the meowing voice, and can be well presented by the INTERSPEECH 2009 emotion challenge feature set. In this subsection, we further analyze the 3 categories of cats' emotion as positive, neutral, and negative with the properties of meowing signals and features.

Frequency Domain Analysis - Spectrum and Spectrogram. Frequency domain properties of cats' meowing voice are displayed in Fig. 6 in form of short time spectrum, from selected typical meowing samples. Similar to human voice, cat voice also presents clear peaks in the spectrum as fundamental frequency and its harmonies. We can see from Fig. 6 that meowing in positive state presents less energy in high frequency band (3000 Hz) than neutral and negative states, and the harmonies are clearer. Meowing in negative state presents a lot of high energy frequencies between the harmony peaks, to make the peak pattern somehow in chaos.

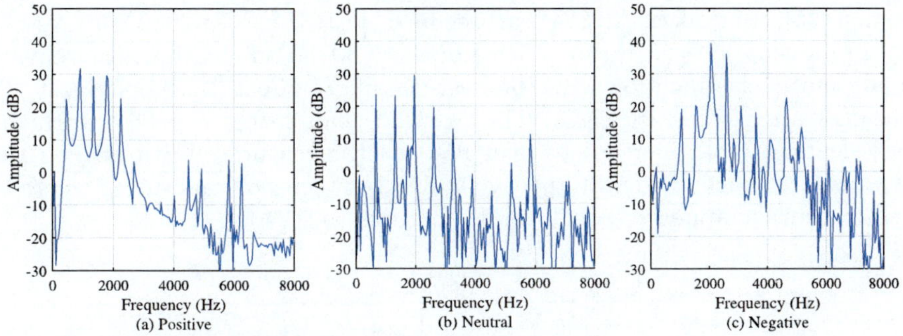

Fig. 6. Spectrum of typical meowing samples of the three categories

A more intuitive illustration of meowing voice can be exhibited in form of spectrogram, as shown in Fig. 7. Cat meowing signals from all 3 emotional states show horizontal stripes in the spectrogram, while the stripes are clearer and thinner in positive state than the other two states, where in neutral states, the stripes are still clear from each other, while in negative state, some of the stripes get blurred together. Neutral state shows smoother F0 trace with the calm voice, positive and negative states that with higher arousal will introduce more fluctuate in F0 trace. Another phenomenon to be noticed is in long scale time domain that, when a cat is in negative state, it tends to produce a longer meowing.

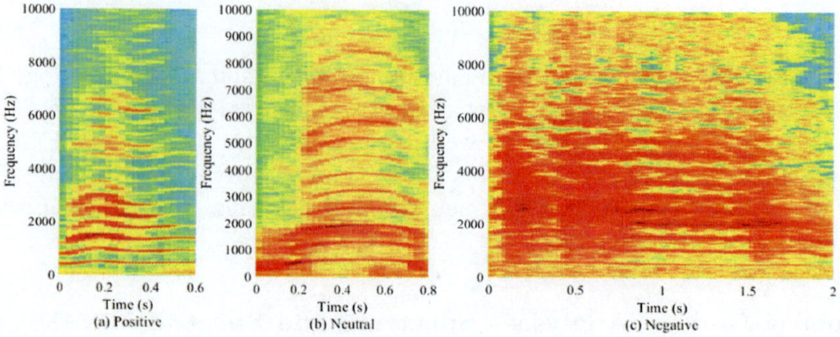

Fig. 7. Spectrogram of typical meowing samples of the three categories

Feature Analysis and Dimension Reduction. This work suffers from a problem that the meowing samples we collected from the internet videos are not sufficiently enough. Even we choose a relatively small scaled feature set, the audio samples in each category are still less than the number of features in the set. Thus, a feature selection, or a feature dimension reduction is necessary for the reliable of this investigation on cats' emotions. In a filter approach of feature selection [25], we ranked the 384 features in INTERSPEECH 2009 feature set in sense of information gain ratio. With from 1 to 30 "good" features, we repeated the automatic classification with the best classifier in Sect. 4.2, LogitBoost. The accuracies are plotted in Fig. 8. The accuracy can reach over 90% with only 4 best features, and increase to over 95% with 8 features. With no less than 10 features, the accuracy stay relative stable between 96% and 97%, where the highest accuracy appears with 17 or 18 features as 97.40%.

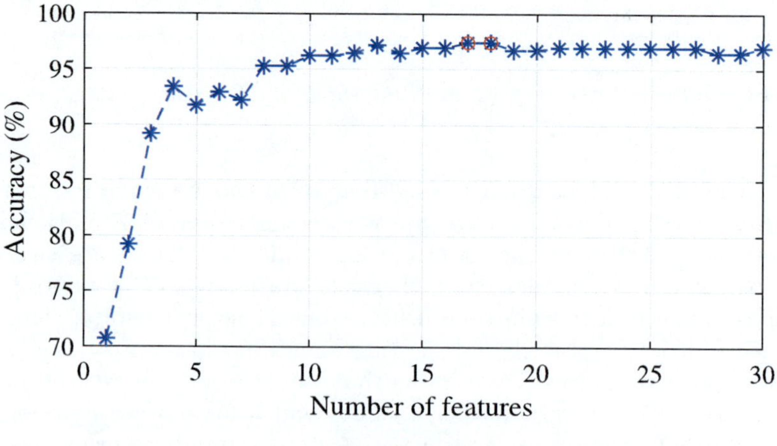

Fig. 8. Accuracy with different number of ranked features

The confusion matrix with the highest accuracy is displayed in Fig. 9. The most significate confusion in this case is that samples of negative state are misjudged as positive at a rate of 4.96%.

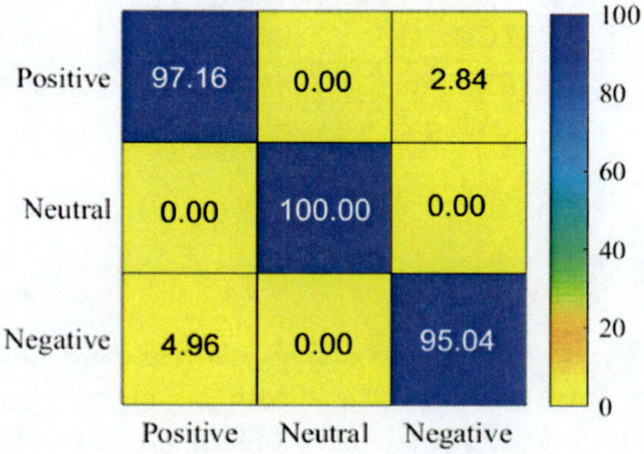

Fig. 9. Confusion matrix with 17 best features

The distribution of 10 selected features is displayed in Fig. 10. As some of the features have similar distribution to each other, these 10 features are not precisely the best 10 features in the ranking. It is shown that these features present different ranges on the three emotional categories, and thus provide distinguishing ability in automatic emotion classification.

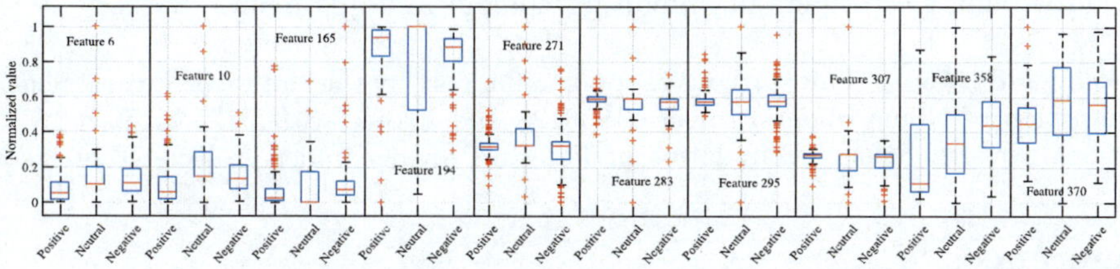

Fig. 10. Distribution of several selected features

The related low level parameters (LLDs as described in INTERSPEECH 2009 feature set) of these features are listed in Table 1. Parameters as RMS energy, ZCR, and MFCC are all important expressive parameters in cats' meowing emotions.

Table 1. Related parameters of the selected features

Feature index	Related LLD	Feature index	Related LLD
6	RMS energy	283	7th MFCC
10	RMS energy	295	8th MFCC
165	ZCR	307	9th MFCC
194	RMS energy	358	ZCR
271	6th MFCC	370	HNR

5 Conclusion

Three categories of cats' emotion, positive, neutral, and negative, as evaluated in valence sense, are investigated with automatic classification on voice signals of cats' meowing. Only audio signals are considered in the learning models, but the ground truth of each sample is determined from video clips with cat voice, facial expression, body movement, as well as their interaction with their owners. Feature set adopted from INTERSPEECH 2009 emotion challenge is proved to be also effective in cats' emotion recognition, and the most expressive features relate to RMS energy, ZCR, and MFCC. The best classification accuracy is obtained from LogitBoost model as 97.40%.

Larger meowing dataset and more detailed emotional categories will be studied in the near future to provide a more accurate and more practical recognition of cats' emotions. Both the aims of this work and the future work focus on the helping of a higher quality human-cat interaction, and make the most of the accompanying role of pets in human psychological adjustment.

Acknowledgment. This work was supported in part by the National Natural Science Foundation of China under Project 61906128 and Project 61802272, in part by the National Natural Science Foundation of Jiangsu Province under project BK20180834.

References

1. Li, S., Yan, Z., Wu, X., Li, A., Zhao, B.: A method of emotional analysis of movie based on convolution neural network and bi-directional LSTM RNN. In: 2017 IEEE Second International Conference on Data Science in Cyberspace (DSC), Shenzhen, pp. 156–161 (2017). https://doi.org/10.1109/DSC.2017.15
2. Xiao, Z., Wu, D., Zhang, X., Tao, Z.: Speech emotion recognition cross language families: Mandarin vs. western languages. In: 2016 International Conference on Progress in Informatics and Computing (PIC), Shanghai, pp. 253–257 (2016). https://doi.org/10.1109/PIC.2016.7949505
3. Kaur, R., Joshi, A.: A study of speech emotion recognition methods. Int. J. Comput. Sci. Mob. Comput. **2** (2013)
4. Xiao, Z., Chen, Y., Dou, W., Tao, Z., Chen, L.: MES-P: an emotional tonal speech dataset in mandarin with distal and proximal labels. IEEE Trans. Affective Comput. (2019). https://doi.org/10.1109/TAFFC.2019.2945322

5. Hantke, S., Cummins, N., Schuller, B.: What is my dog trying to tell me? the automatic recognition of the context and perceived emotion of dog barks. In: 2018 IEEE International Conference on Acoustics, Speech and Signal Processing (ICASSP), Calgary, AB, pp. 5134–5138 (2018). https://doi.org/10.1109/ICASSP.2018.8461757
6. Ekman, P.: An argument for basic emotions. In: Cognition and Emotion, vol. 6 (1992). https://doi.org/10.1080/02699939208411068
7. Scherer, K.: Psychological models of emotion. In: The Neuropsychology of Emotion (2000)
8. Engberg, I.S., Hansen, A.V., Andersen, O.K., Dalsgaard, P.: Design, recording and verification of a Danish emotional speech database. In: European Conference on Speech Communication and Technology, Rhodes, Greece (1997)
9. Burkhardt, F., Paeschke, A., Rolfes, M., Sendlmeier, W.F., Weiss, B.: A database of German emotional speech. In: INTERSPEECH 2005 - Eurospeech, 9th European Conference on Speech Communication and Technology, Lisbon, Portugal (2005)
10. Busso, C., Parthasarathy, S., Burmania, A., Abdelwahab, M., Sadoughi, N., Provost, E.M.: MSP-IMPROV: an acted corpus of dyadic interactions to study emotion perception. IEEE Trans. Affect. Comput. **8**, 67–80 (2017). https://doi.org/10.1109/TAFFC.2016.2515617
11. Lakatos, G.: Dogs as behavior models for companion robots: how can Human-Dog interactions assist social robotics? IEEE Trans. Cogn. Dev. Syst. **9**, 234–240 (2017). https://doi.org/10.1109/TCDS.2016.2552244
12. Lakatos, G.: Dogs as behavior models for companion robots: How can Human-Dog interactions assist social robotics? IEEE Trans. Cogn. Dev. Syst. **9**, 234–240 (2017). https://doi.org/10.1109/TCDS.2016.2552244
13. Molnár, C.: Classification of Dog barks: a machine learning approach. Animal Cogn. **11**, 389–400 (2008). https://doi.org/10.1007/s10071-007-0129-9
14. Pongracz, P., Molnar, C., Miklosi, A., Csányi, V.: Human listeners are able to classify dog (Canis familiaris) barks recorded in different situations. J. Comp. Psychol. 119, 136. Washington, D.C (2005). https://doi.org/10.1037/0735-7036.119.2.136
15. Molnár, C., Pongrácz, P., Dóka, A., Miklósi, A.: Can humans discriminate between dogs on the base of the acoustic parameters of barks? Behav. Process/ **73**, 76–83 (2006). https://doi.org/10.1016/j.beproc.2006.03.014
16. Pongrácz, P., Miklósi, D., Csányi, V.: Owner's beliefs on the ability of their pet dogs to understand human verbal communication: a case of social understanding. Curr. Psychol. Cogn. (2000)
17. Faragó, T., Takács, N., Miklósi, A., Pongracz, P.: Dog growls express various contextual and affective content for human listeners. R. Soc. Open Sci. **4** (2017). https://doi.org/10.1098/rsos.170134. England
18. Schuller, B., Steidl, S., Batliner, A.: The interspeech 2009 emotion challenge. In: Proceedings of Interspeech, pp. 312–315 (2009)
19. Schuller, B., Steidl, S., Batliner, A., Burkhardt, F., Devillers, F., Muller, C., Narayanan, S.: The interspeech 2010 paralinguistic challenge. In: Proceedings of the 11th Annual Conference of the International Speech Communication Association, pp. 2794–2797 (2010)
20. Schuller, B., et al.: The interspeech 2013 computational paralinguistics challenge: social signals, conflict, emotion, autism. In: Proceedings of the Annual Conference of the International Speech Communication Association, INTERSPEECH, Lyon, France, pp. 148–152 (2013)

21. Deb, S., Dandapat, S., Krajewski, J.: Analysis and classification of cold speech using variational mode decomposition. IEEE Trans. Affect. Comput. **11**, 296–307 (2020). https://doi.org/10.1109/TAFFC.2017.2761750
22. Eyben, F., Wollmer, M., Schuller, B.: Opensmile - the munich versatile and fast open-source audio feature extractor. In: ACM MM, pp. 1459–1462 (2010). https://doi.org/10.1145/1873951.1874246
23. Hall, M.A., Frank, E., Holmes, G., Pfahringer, B., Reutemann, P., Witten, I.H.: The WEKA data mining software: an update. SIGKDD Explor. **11**, 10–18 (2008). https://doi.org/10.1145/1656274.1656278
24. He, H., Garcia, E.A.: Learning from imbalanced data. IEEE Trans. Knowl. Data Eng. **21**, 1263–1284 (2009). https://doi.org/10.1109/TKDE.2008.239
25. Kojadinovic, I., Wottka, T.: Comparison between a filter and a wrapper approach to variable subsetselection in regression problems (2000)

Computer Audition for Healthcare

Are You Speaking with a Mask? An Investigation on Attention Based Deep Temporal Convolutional Neural Networks for Mask Detection Task

Yu Qiao[1], Kun Qian[2(✉)], Ziping Zhao[1(✉)], and Xiaojing Zhao[1]

[1] Tianjin Normal University, Tianjin, China
[2] The University of Tokyo, Tokyo, Japan
qian@p.u-tokyo.ac.jp

Abstract. When writing this article, COVID-19 as a global epidemic, has affected more than 200 countries and territories globally and lead to more than 694,000 deaths. Wearing a mask is one of most convenient, cheap, and efficient precautions. Moreover, guaranteeing a quality of the speech under the condition of wearing a mask is crucial in real-world telecommunication technologies. To this line, the goal of the ComParE 2020 Mask condition recognition of speakers subchallenge is to recognize the states of speakers with or without facial masks worn. In this work, we present three modeling methods under the deep neural network framework, namely Convolutional Recurrent Neural Network(CRNN), Convolutional Temporal Convolutional Network(CTCNs) and CTCNs combined with utterance level features, respectively. Furthermore, we use cycle mode to fill the samples to further enhance the system performance. In the CTCNs model, we tried different network depths. Finally, the experimental results demonstrate the effectiveness of the CTCNs network structure, which can reach an unweighted average recall (UAR) at 66.4% on the development set. This is higher than the result of baseline, which is 64.4% in S2SAE+SVM nerwork(a significance level at $p < 0.001$ by one-tailed z-test). It demonstrates the good performance of our proposed network.

Keywords: Computational paralinguistics · Deep learning framework · Mask condition recognition · Speech recognition

1 Introduction

COVID-19, as a pandemic, has more than 20 million confirmed patients (causing more than 748 000 deaths), and is still affecting more than 200 countries and territories globally at the time of writing this paper[1]. Computer audition (CA), a multidisciplinary field that leverages the advanced acoustic/audio signal processing and machine learning technologies to enable the machines having or even

[1] https://coronavirus.jhu.edu/map.html.

X. Shao et al. (Eds.): CSMT 2020, LNEE 761, pp. 163–174, 2021.
https://doi.org/10.1007/978-981-16-1649-5_14

outperforming the human hearing capacities, has been increasingly applied to the healthcare domain [12]. More recently, CA has been thought to have promising potential for fighting the COVID-19 pandemic due to its non-invasive and ubiquitous characteristic by nature [9,15].

In this paper, we aim to develop a speech-driven deep learning framework to recognize people with or without facial masks worn. The task is proposed as part of the INTERSPEECH 2020 Computational Paralinguistics ChallengE (ComParE) [14]. The data offered in this challenge is called the MASC (the Mask Augsburg Speech Corpus) dataset, which is the first to give access to recordings of speech from individuals wearing an operation mask. The labels of the data are their condition states while communicating, including Masking and Clear. Many existing works have been performed on the speech recognition research. Some acoustic features, such as the extended Geneva Minimalistic Acoustic Parameter Set (eGeMAPS) [7], ComParE acoustic feature set and Bag of-Audio-Words (BoAW) feature set [13], combined with traditional machine learning methods, have been proved to be effective for recognizing the speech signals.

With the development of deep learning, neural network has made substantial achievements in the computational paralinguistics field. Neural network have been widely used due to the superior performance, such as speaker identification [11,17,19,21], language identification [1–3] and speech emotion recognition [5,20]. Therefore, various neural network frameworks such as convolutional neural network (CNN) and recursive neural network (RNN) have emerged. CNN is used to extract spatial features and generate feature maps. The extensive application from AlexNet to VGG model reflects the superior performance of CNN. The pre-trained AlexNet network was used to extract deep features, and then the features of the full connection layer were input into Support Vector Machine (SVM) for classification, which achieved good performance on the data set FAU-AIBO [6]. Two different convolution nuclei were used to extract time-domain and frequency-domain features respectively, and then the features were classified by CNN after fusion. Finally, the UAR of the four categories of emotions of IEMOCAP reached 68% [10]. Gated Recurrent Unit (GRU) and Long-Short Term Memory (LSTM) are also widely used, GRU is a variant of LSTM, they can solve the gradient vanishing problem in the RNN optimization process. Greff et al. benchmarked eight LSTM variants on speech recognition [8]. The combination of CNN and RNN is widely used. Mingyi et al. added LSTM after CNN, and found that the five convolution layers had the best performance on EmoDB [4]. However, with the deepening of network layers, some information will be lost because CNN has no memory function, and the operation time of RNN is relatively long.

Main contributions of this work can be summarised as follows: First, we compare the performance of two different network topologies on this classification problem and find the good effect of TCN on this classification problem. Second, we have introduced attention mechanism across all network structures to allow the network to focus on key features during training. Third, we integrate utterance level features into the network structure with good performance, realized the fusion of deep learning representation and utterance level features.

In this article, We investigate and compare three topologies, i.e., Convolutional Recurrent NeuralNetwork (CRNN), Convolutional Temporal Convolutional Networ (CTCNs) and CTCNs with utterance level features. In addition, CNN and attention are added to both models to improve the network performance.

This paper is organized as follows: Firstly, we introduce the methods used in Sect. 2. Section 3 introduces experimental design, including data preprocessing, experimental setting, and experimental results. And the discussion will be given in Sect. 4. Finally, we conclude this study in Sect. 5.

2 Methods

2.1 BLSTM

BLSTM is composed of forward LSTM and backward LSTM. In the LSTM, there are three kinds of gates: forgetting gate, input gate and output gate. The forgetting gate can selectively forget some information, and the input gate new information selectively recorded, and in the output gate for output. In BLSTM, forward LSTM is used to help the network learn sequence characteristics forward and backward LSTM learns sequence information later. This design can help the network form sequence memory. When we input the extracted mask audio sequence, we can not only accumulate the information of the input moment, but also remember the information of the previous moment, which has a good effect on dealing with the time series problem.

The network structure diagram of BLSTM is shown in Fig. 1, from which we can see that the output layers results are jointly controlled by forward layers and backword layers, and the final output results can be expressed as follows by mathematical expressions:

$$h_t = f(w_1 x_t + w_2 h_{t-1}) \tag{1}$$

$$h_t^{'} = f(w_4 x_t + w_5 h_{t+1}^{'}) \tag{2}$$

$$O_t = g(w_3 h_t + w_6 h_t^{'}) \tag{3}$$

where, Eq. (1) represents the result of forward propagation, Eq. (2) represents the result of back propagation, and Eq. (3) represents the expression of the output result after BLSTM.

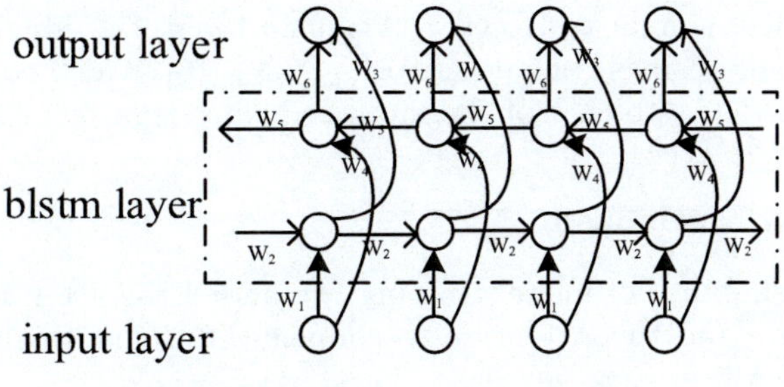

Fig. 1. BLSTM network structure.

2.2 TCN

Similar to BLSTM, TCN can also be used to handle time series problems. TCN network is all convolution operation, which means that TCN neural network can carry out large-scale parallel processing, which is shorter than BLSTM to some extent, which involves the skip layer connection of dilated convolution, causal convolution, and residual convolution.

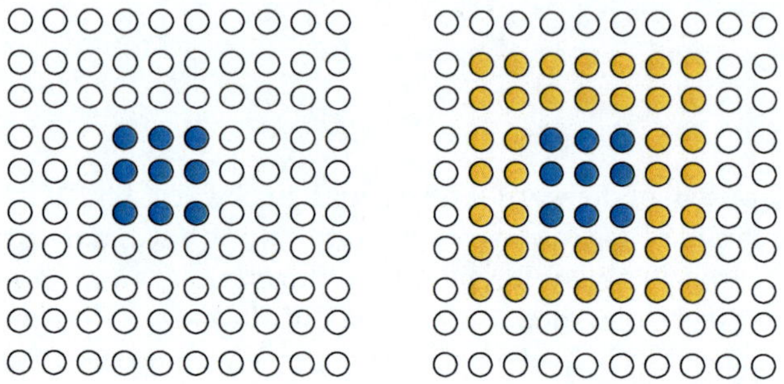

(a) General Convolution (b) Dilated Convolution

Fig. 2. Contrast diagram of convolution receptive field.

Dilation rate parameter is involved in the part of dilation convolution, which is used to represent the size of the dilation, so that the convolution process has a larger receptive field. As shown in Fig. 2(a) represents the receptive field of dilated convolution, and (b) represents the receptive field of general convolution. From the figure, the advantages of the receptive field of dilated convolution can be clearly seen. Where, the size of the convolution kernel in (a) is 3, and after dilation rate, the size of the convolution kernel becomes 5, and finally the receptive field of (b) is obtained.

Where, the calculation of the size of the dilated convolution kernel follows: dilated filter $= d * (k - 1) + 1$, where d stands for dilation rate and k stands for the size of the convolution kernel.

By referring the dilative convolution to the causal convolution, the prediction at time t can take into account the sequence before time$t$, thus achieving a time memory effect similar to BLSTM. The skip layer of residual convolution is realized by 1D fully-convolutional network (FCN) [16], which equals the length of the output sequence to that of the input sequence [22].

2.3 Attention Mechanism

To ensure the reliability of model training, we added the attention layer to the network structure and the Attention mechanism after the weight causal layers in TCN, as shown in Fig. 3.

In this paper, the attention layer in the network structure is a sequence coding layer, which is a series of weight allocation coefficients. When the input information at time t is more similar to the target information, the attention layer assigns more weight to the time t, that is, the output of the sequence is more dependent on the time t. In the experiment of this paper, Self-Attention mechanism is used, which can find the internal connection of the sequence in the training process, so as to ensure the similarity between the output sequence and the input sequence. So, we use Scaled Dot-Product Attention [18], the implementation equation is

$$Attention(Q, K, V) = softmax(\frac{Q \times K^T}{\sqrt{d_k}}) \times V \tag{4}$$

Here, K and V are the values of mask audio data after Self-Attention, Q is the data that corresponds to the label by masked Self - Attention after the value, d_k is the number of channels in the input sequence, used as a normalization.

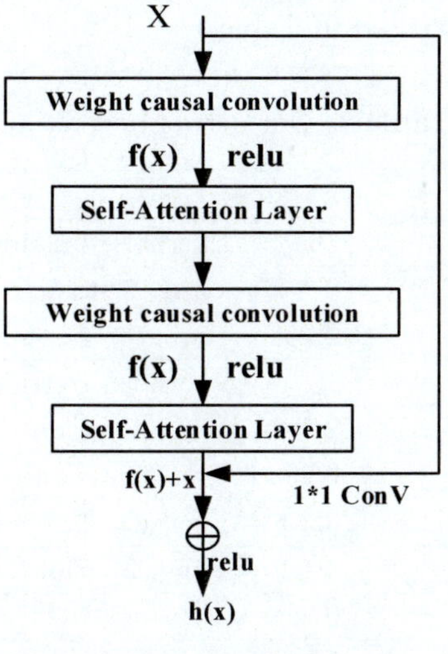

Fig. 3. Attention residual learning block.

3 Experiment Design

3.1 Data Pre-processing

In this part, all the audio data in the data set are circulated and filled in for 4s (the original data set lasts for 1s and the sampling frequency is 16 kHZ). Then, the librosa library is used to perform short-time Fourier transform to extract mel spectrogram. The parameters in the process of mel spectrogram extraction are as follows: the window width w = 25 ms, the window shift 10 ms, and n_{mels} = 128 mel frequency bands.

3.2 Experimental Setting

In our experiment, we mainly used three network learning models: CRNN, CTCNs and CTCNs with utterance level features. We will describe these three network structures in detail below. It should be noted that due to the limitation of server storage space, the batchsize of all our experiments is 64.

CRNN. In this model, we first used CNNs to extract features from the mel spectrograms, considering the effect of the preceding sequence on the prediction of the following sequence, we use BLSTM to remember information through forward propagation and backward propagation, so as to make the predicted results more robust. At the same time, after BLSTM layer, add the attention layer to allocate the feature weight, so that the network can focus on the features that play a key role in the classification effect. Finally, the spatial features extracted from the convolutional layer and the sequence features after the attention layer are fused as the final classification features, and the classification is carried out through the full connection layer containing softmax function. More specifically, our network model is described in Table 1.

Table 1. Our network structure

Nework layers	Parameter
Conv1	16, 7 * 7 kernels, 1 stride
Pooling	2 * 2 pooling, 2 stride
Dropout	0.25
Conv2	16, 5 * 5 kernels, 1 stride
Pooling	2 * 2 pooling, 2 stride
Dropout	0.25
Conv3	32, 5 * 5 kernels, 1 stride
Pooling	2 * 2 pooling, 2 stride
Dropout	0.25
MaxPooling	BLSTM/
	TCN blocks: 3 * 3 kernels,d: [1, 2, 4...]
	Self-attention layer
Features concatenation	
Full-connected Layer	4096 units
Classification Layers	Softmax

CTCNs. In this network structure, we use TCN and attention layer to build the network structure. In the TCN module, we mainly used the structure in Fig. 3. Multi-layer stacking was performed in residual block in Fig. 3 to build the main part of the network. Of course, with the stack of blocks, the number of layers in the network would be deepened, and the attention layer would be added after the last layer. This approach is to achieve similar functions to BLSTM, enabling the network to extract time series features. Considering the impact of spatial features on the classification results, we added three convolutional layers at the beginning of the network. The features extracted by the convolutional layer were on the one hand input into the TCN network module, and on the other hand retained and fused with the sequence features extracted by the TCN module, thus forming the features of final progressive classification. The final features are sorted through the full connection layer of 4096 units by softmax. The detailed network structure is shown in Table 1.

CTCNs with Utterance Level Features. In the experiment, we mainly used the manually designed features of low level descriptors (LLDs) and high level statistics functions (HSFs), obtained utterance level features by making statistics on the voice features at the frame level, such as maximum value and mean value, and so on. Here, opensmile toolkit is used to extract utterance level features, and the feature Set used is ComParE.

In this part, we added utterance level features to integrate the deep features extracted from deep learning for classification. The extraction of deep features is based on the experiment in Sect. 3.2, and the features extracted from its full connection layer are used.

Refer to the specific network structure Fig. 4.

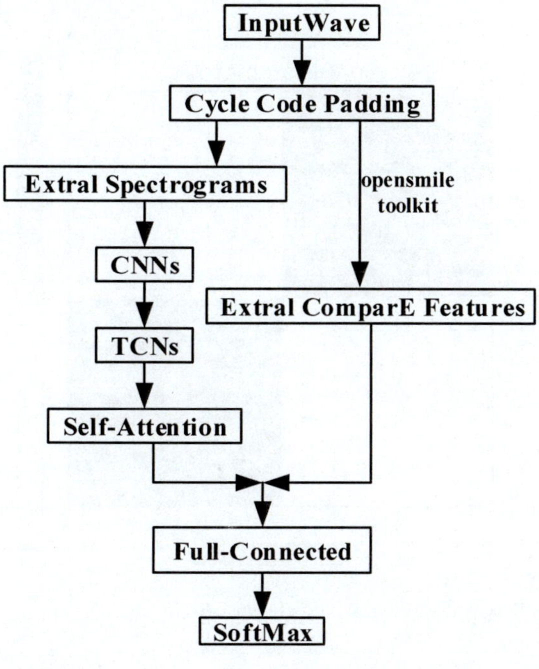

Fig. 4. TCNs with utterance level features network structure.

Experimental Result. In this paper, we will use unweighted average recall (UAR) to evaluate the experimental results of various network structures. As this is a Sub-Challenge task, all our results are obtained on the development set. For Sect. 3.2, we conducted experiments with [3,10] different attention residual blocks, and the experimental results are shown in Table 2. It can be seen from the table that when residual blocks is 4, the experimental UAR is 66.4%, which is the best result. The number of channels and experiment time for each block are also shown in Table 2. The confusion matrix corresponding to the experiment is shown in the Fig. 5.

Table 2. The result of CTCNs network structure on the development set

Nework blocks	Channels	WAR (%)	UAR (%)	Time (s)
3	[64, 128, 256]	64.6	65.1	3692.98
4	[64, 128, 256, 512]	66.3	**66.4**	4874.648
5	[64, 128, 256, 512, 1024]	64.7	64.8	4218.848
6	[64, 128, 256, 512, 1024, 2048]	66.3	65.6	8350.363
7	[64, 128, 256, 512, 1024, 2048, 4096]	65.4	65.0	12205.034
8	[64, 64, 128, 128, 256, 256, 512, 1024]	66.6	66.0	10779.831
9	[64, 64, 64, 128, 128, 128, 256, 256, 512]	64.8	64.4	5451.703
10	[64, 64, 64, 128, 128, 128, 256, 256, 512, 1024]	65.2	65.1	15232.913

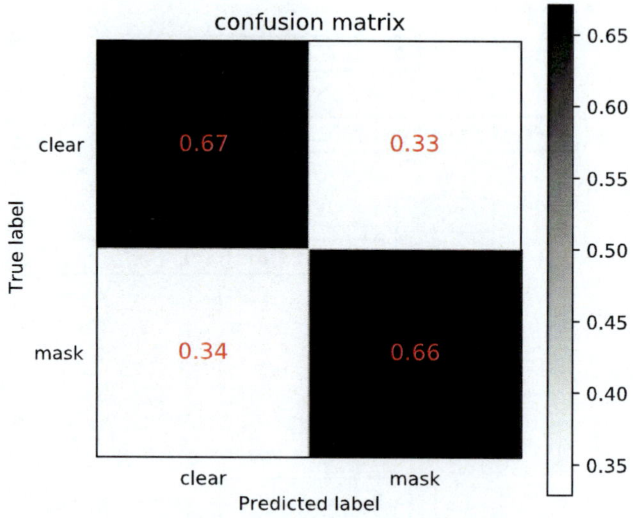

Fig. 5. Confusion matrix graph on the development set.

Table 3. Results of different network structures on the development set

ID	Nework structure	UAR(%)
1	CRNN	65.5
2	CTCNs	**66.4**
3	CTCNs + ComparE	65.9
4	ComparE + SVM [14]	62.6
5	ComparE BOAW + SVM [14]	64.2
6	DeepSpectrum + SVM [14]	63.4
7	S2SAE + SVM [14]	64.4

As can be seen from the table, the lowest experimental result of our proposed method is 65.5%, while the experimental result of S2SAE model in the original paper is the best, with its UAR being 64.4%, which is lower than our lowest result by 1.1%, which fully proves the performance of our network structure.

4 Discussion

It can be seen from Table 2 that the network of 4-layer blocks has the best result on the development set. As the network deepens to 10 layer blocks, the UAR of the network is not as good as that of 4-layer blocks. This may be from the side that the deepening of the network makes the training gradient unstable. In Table 2, we can see that when blocks is 7 or 8, channels are the most and the experiment takes more than 10,000 s.

The experimental results of different network structures are shown in Table 3. The model 1, 2, 3 network structures are the three methods tried in this paper, and the model 4,5,6,7 are the experimental results of the original paper's network structures. The difference between model 2 and model 1 is that model 2 uses TCN to extract sequence features, while model 1 uses BLSTM, and it is finally found that the experimental results of model 2 are better than those of model 1, which maybe indicates that TCN has a better fitting on this data set. When we fusioned utterance level features (in this article, ComParE the features) into model 2, the experimental result is 65.9% in model 3, but this reduced the results by 0.5%. We consider the reasons for this result may be to join utterance level features, making increased certain features of the similarity between different categories, it increases the classification error, thus resulting in a loss of the experimental results. It may be possible to try other utterance level features for fusion, hoping to improve the classification result. Model 3 is about 3% higher than model 4, and it turns out that the TCN network extracts features that are useful for classification.

5 Conclusion

Mask Sub-Challenge detection is a challenging task. In this paper, we first adopted the cycle code padding method to process the raw audio, and then conducted experiments on the MASC data set through three different network structures, namely CRNN, CTCNs and CTCNs with utterance level features. CTCNs achieves the best performance on the development set.

The experimental result of model 4 is the lowest, which used only ComParE features, while model 2 adds spectral features on this basis, the results increased by 3.3%, which may indicate the advantage of mel spectrograms in this data set. All the deep feature extraction in this paper is based on the spectrograms extracted by the short-time Fourier Transform (STFT). However, window size in the process of STFT do not have adaptivity and cannot be optimized for specific problems, so better results may be obtained by using wavelet transform to extract spectrograms.

The experimental results of other models are better than model 4, which may reflect the good performance of deep learning. This suggests that we should not be confined to machine learning, and future research can be developed towards deep learning, perhaps with better results.

Acknowledgements. This work was partially supported by the National Natural Science Foundation of China (Grant No. 61702370), P. R. China, the Key Program of the Natural Science Foundation of Tianjin (Grant No. 18JCZDJC36300), P. R. China, the Open Projects Program of the National Laboratory of Pattern Recognition, P. R. China, the Zhejiang Lab's International Talent Fund for Young Professionals (Project HANAMI), P. R. China, the JSPS Postdoctoral Fellowship for Research in Japan (ID No. P19081) from the Japan Society for the Promotion of Science (JSPS), Japan, and the Grants-in-Aid for Scientific Research (No. 19F19081) from the Ministry of Education, Culture, Sports, Science and Technology, Japan.

References

1. Bartz, C., Herold, T., Yang, H., Meinel, C.: Language identification using deep convolutional recurrent neural networks. In: Proceedings of the 24th International Conference of Neural Information Processing, pp. 880–889. Springer, Guangzhou, China (2017)
2. Cai, W., Cai, D., Huang, S., Li, M.: Utterance-level end-to-end language identification using attention-based cnn-blstm. In: Proceedings of the 2019 IEEE International Conference on Acoustics, Speech and Signal Processing (ICASSP), pp. 1–5. IEEE, Brighton, UK (2019)
3. Chan, W., Jaitly, N., Le, Q., Vinyals, O.: Listen, attend and spell: a neural network for large vocabulary conversational speech recognition. In: Proceedings of the 2016 IEEE International Conference on Acoustics, Speech and Signal Processing (ICASSP), pp. 4960–4964. IEEE, Shanghai, China (2016)
4. Chen, M., He, X., Yang, J., Zhang, H.: 3-D convolutional recurrent neural networks with attention model for speech emotion recognition. IEEE Sig. Process. Lett. **25**(10), 1440–1444 (2018)

5. Chernykh, V., Sterling, G., Prihodko, P.: Emotion recognition from speech with recurrent neural networks, pp.1–18 (2017). ArXiv:abs/1701.08071
6. Cummins, N., Amiriparian, S., Hagerer, G., Batliner, A., Steidl, S., Schuller, B.W.: An image-based deep spectrum feature representation for the recognition of emotional speech. In: Proceedings of the 25th ACM International Conference on Multimedia, pp. 478–484. Association for Computing Machinery, Seattle, USA (2017)
7. Eyben, F.: The Geneva minimalistic acoustic parameter set (GeMAPS) for voice research and affective computing. IEEE Trans. Affect. Comput. **7**(2), 190–202 (2016)
8. Greff, K., Srivastava, R.K., Koutník, J., Steunebrink, B.R., Schmidhuber, J.: LSTM: a search space odyssey. IEEE Trans. Neural Netw. Learn. Syst. **28**(10), 2222–2232 (2017)
9. Han, J., Qian, K., Song, M., Yang, Z., Ren, Z., Liu, S., Liu, J., Zheng, H., Ji, W., Koike, T., et al.: An early study on intelligent analysis of speech under Covid-19: Severity, sleep quality, fatigue, and anxiety. In: Proceedings of Interspeech, pp. 4946–4950. Shanghai, China (2020)
10. Li, P., Song, Y., McLoughlin, I.V., Guo, W., Dai, L.R.: An attention pooling based representation learning method for speech emotion recognition. In: Proceedings of Interspeech. ISCA, Hyderabad, India, pp. 3087–3091 (2018)
11. Matějka, P., Glembek, O., Novotny, O., Plchot, O., Grézl, F., Burget, L., Cernocky, J.: Analysis of dnn approaches to speaker identification. In: Proceedings of the 41st IEEE International Conference on Acoustics, Speech and Signal Processing (ICASSP), pp. 5100–5104. IEEE, Shanghai, China (2016)
12. Qian, K., Li, X., Li, H., Li, S., Li, W., Ning, Z., Yu, S., Hou, L., Tang, G., Lu, J., Li, F., Duan, S., Du, C., Cheng, Y., Wang, Y., Gan, L., Yamamoto, Y., Schuller, B.W.: Computer audition for healthcare: opportunities and challenges. Front. Digit. Health **2**, 5 (2020)
13. Schmitt, M., Schuller, B.: openXBOW - introducing the Passau open-source cross-modal bag-of-words toolkit. J. Mach. Learn. Res. **18**(96), 1–5 (2017)
14. Schuller, B.W., et al.: The INTERSPEECH 2020 Computational Paralinguistics Challenge: Elderly emotion, Breathing & Masks. In: Proceedings of Interspeech, pp. 2042–2046. Shanghai, China (2020)
15. Schuller, B.W., Schuller, D.M., Qian, K., Liu, J., Zheng, H., Li, X.: Covid-19 and computer audition: an overview on what speech & sound analysis could contribute in the SARS-CoV-2 corona crisis, pp. 1–7. arXiv preprint arXiv:2003.11117 (2020)
16. Shelhamer, E., Long, J., Darrell, T.: Fully convolutional networks for semantic segmentation. IEEE Trans. Pattern Anal. Mach. Intell. **39**(4), 640–651 (2017)
17. Snyder, D., Ghahremani, P., Povey, D., Garcia-Romero, D., Carmiel, Y., Khudanpur, S.: Deep neural network-based speaker embeddings for end-to-end speaker verification. In: Proceedings of the 2016 IEEE Spoken Language Technology Workshop (SLT), pp. 165–170. IEEE, San Juan, Puerto Rico (2016)
18. Vaswani, A., et al.: Attention is all you need. In: Proceedings of the 31st International Conference on Neural Information Processing Systems (NIPS), pp. 5998–6008. Curran Associates Inc., Long Beach, CA, USA (2017)
19. Villalba, J., Brümmer, N., Dehak, N.: Tied variational autoencoder backends for i-vector speaker recognition. In: Proceedings of Interspeech, pp. 1004–1008. ISCA, Stockholm, Sweden (2017)
20. Xie, J., Xu, X., Shu, L.: WT feature based emotion recognition from multi-channel physiological signals with decision fusion. In: Proceedings of the 2018 First Asian Conference on Affective Computing and Intelligent Interaction (ACII Asia), pp. 1–6. IEEE, Beijing, China (2018)

21. Xie, W., Nagrani, A., Chung, J.S., Zisserman, A.: Utterance-level aggregation for speaker recognition in the wild. In: Proceedings of the International Conference on Acoustics, Speech, and Signal Processing (ICASSP), pp. 5791–5795. IEEE, Brighton, UK (2019)
22. Yu, F., Koltun, V., Funkhouser, T.A.: Dilated residual networks. In: Proceedings of 2017 IEEE Conference on Computer Vision and Pattern Recognition (CVPR), pp. 636–644. IEEE, Honolulu, Hawaii (2017)

CSMT 2020 Challenge Papers

A Novel Dataset for the Identification of Computer Generated Melodies in the CSMT Challenge

Shengchen Li[1(✉)], Yinji Jing[2], and György Fazekas[3]

[1] Department of Intelligent Science, School of Advanced Technology, Xi'an Jiaotong-Liverpool University, 111 Ren'ai Road, Suzhou Industrial Park, Suzhou 215123, Jiangsu Province, P. R. China
shengchen.li@xjtlu.edu.cn
[2] Beijing University of Posts and Telecommunications, Beijing, China
jyj@bupt.edu.cn
[3] Queen Mary University of London, London, UK
g.fezekas@qmul.ac.uk

Abstract. This paper introduces a novel dataset for the identification of computer generated melodies as used in the data challenge organised by the Conference on Sound and Music Technology (CSMT). The CSMT data challenge requires participants to identify whether a given piece of melody is generated by computer or is composed by human. The dataset consists of two parts: a development dataset and an evaluation dataset. The development dataset contains only computer generated melodies whereas the evaluation dataset contain both computer generated melodies and human composed melodies. The aim of the dataset is to facilitate the develpment and assessment of methods to identified computer generated melodies and facilitate the creation of generative music systems.

Keywords: Melody clustering · Dataset · Computer generated melody identification

1 Introduction

Automatic music generation is becoming more and more popular with the development of deep learning techniques. At the same time, new challenges have emerged in juridical practices regarding copyright protection: the source of music leads to different juridical models. Although the discussion of legal issues is beyond the scope of this paper and it isn't among the aims of the proposed data challenge, a new task is considered helpful in future juridical practices. Identifying whether a piece of melody is computer generated or human composed could help in recognising cases of music use where legal intervention of further scrutiny is necessary. As a result, the Conference on Sound and Music Technology (CSMT) proposes a data challenge that requires participants to identify human composed melodies among computer generated ones.

Supported by Zhongwen Law Firm. Shengchen Li and Yinji Jing are considered joint first authors.

X. Shao et al. (Eds.): CSMT 2020, LNEE 761, pp. 177–186, 2021.
https://doi.org/10.1007/978-981-16-1649-5_15

Existing automatic music generation methods have certain drawbacks, such as the lack of clear long term structure in the music or the existence of unusual harmonisation, which make the melody identification less challenging. For example, a Self Similarity Matrix (SSM) is used to identify the repetitions in music [4] that are commonly associated to music structure by composers but are seldom present in pieces produced by music generation algorithms [8]. Moreover in juridical practices, copyright infringement can be detected using the similarity of the variation of pitch in melodies, regardless of music structure and accompaniment. In this paper and the proposed challenge, the term "melody" refers to a sequence of pitches with dedicated duration but excludes the concept of music structure and accompaniment.

The proposed data challenge follows a possible scenario of melody source identification in juridical practice. There are two datasets used in the challenge. The development dataset consists of computer-generated melodies that are produced by a set of exemplar music generation systems. The evaluation dataset contains both computer-generated and human-composed melodies. Participants are required to submit a system that identifies human-composed melodies among the computer-generated melodies.

The authors and organisers of the data challenge reviewed existing computer music generation systems as outlined in this paper. Three exemplar methodologies were proposed including Generative Adversarial Network (GAN), Variational Auto Encoder (VAE) and transformer systems because these architectures are commonly used and represent the current state-of-the-art in music generation as of early 2020. All systems were used to produce computer-generated melodies in both development and evaluation datasets. The systems used to generate melodies for development and evaluation datasets are different as the results of a different initial values and different batch formation in the training process. For human-composed melodies in the evaluation dataset, the majority (95%) of human-composed melodies within the evaluation dataset overlaps with human-composed melodies used as training data for the automatic generation system. The remaining human-composed melodies are composed by university students whose major is music composition. Such melodies have not been published to the public.

The proposed data challenge can be approached in two different ways. If human-composed melodies are collected by the participants, data may be labelled as "human" vs. "computer", hence the proposed task can be considered a binary classification problem. The human-composed melodies can also be considered outliers among computer-generated melodies. In this case, the proposed task can also be viewed as an unsupervised outlier detection problem.

The rest of the paper is organised as follows. A brief overview of automatic music generation is presented in Sect. 2 in order to justify the choice of melody generation systems. In Sect. 3, the dataset creation process is explained in detail together with the data representation proposed for the challenge. This is followed by a brief conclusion in Sect. 4.

2 Melody Generation Systems

This section provides an overview of automatic music generation systems. The majority of music generation systems can be divided to three types [10]: rule-based systems, methods that utilise mathematical models and machine learning systems. The machine learning systems, especially deep learning systems, are considered as the state-of-the-art automatic music generation systems [2]. As a result, the data challenge proposes to use deep learning systems to generate melodies that are labelled as computer-generated melodies.

The most important factor for automatic music generation that affects system performance is the modelling of temporal dependencies. Rule-based systems usually propose a set of rules to generate a sequences such as chords [16]. Systems that use mathematical models aim to describe time dependency in music mathematically. The generation process may then be considered a sampling process from a mathematical model. For modelling temporal dependencies, Markov models are considered the first choice since the very early stages of music generation [14]. One of more the recent works using this principle is the ALYSIA system [1] that creates both lyrics and melodies.

As music usually has a long-time dependency, it is almost impossible for rule-based and mathematical modelling systems to learn long-time dependencies accurately. Machine learning systems especially deep learning systems are better suited for the purpose of music generation as the long-term dependency can be modelled as a joint probability distribution akin to a language model [6].

One exemplar system is a Recurrent Neural Network (RNN). Makris [13] uses RNN to generate rhythm in drum patterns. The Microsoft team [20] uses RNN to encode the pitch, rhythm and chord of music. With the development of transformer systems that are better at modelling longer-time dependencies, Vaswani et al. [17] proposed transformer structure to catch longer temporal-dependency. This was adopted by Huang et al. [9] for music generation. In the proposed data challenge, the MusicTransformer [9] system is used as one of the candidate system to generate computer-generated melodies, where the authors claimed that the MusicTransformer models long-term dependencies in music [9].

Besides using a language model to model long-time dependency in music, music generation can also be performed by a generative model such as a Variational Auto-Encoder (VAE) or a Generative Adversarial Network (GAN).

VAE is a variant of the autoencoder, which is a generative deep learning model. Brunner [3] proposed a VAE-based automatic composition model MIDI-VAE, which processes polyphonic music with multiple instrument tracks and models the duration and speed of the notes in the generated music. Wang [18] proposed a new variant of Variational Autoencoder (VAE), which uses a modular approach to designing the model structure to generate music. Luo [12] used a variational autoencoder to generate different styles of Chinese folk music. MusicVAE [15] improves the structure of VAE according to the characteristics of music with hierarchical structure, which aims to solve the lack of coherence in generated music using vanilla VAE. The MusicVAE system is better at generating music with extended duration hence the proposed data challenge selects MusicVAE as the representative of VAE-based music generation systems in the development and evalution datasets.

Generative adversarial network (GAN) [7] is a generative model that contains a generator and a discriminator. In a GAN, the generator produces pseudo-samples and the discriminator judges whether a sample was produced by the generator. GAN is commonly used for music generation, for example, by Liu and Yang [11] and Dong et al. [5]. MidiNet [19] is one of few GAN systems that use piano roll as the representation of music and can generate melodies without the generation of music accompaniment. As a result, the proposed data challenge selects MidiNet as the choice of GAN based systems for music generation.

To summarise, deep learning based computer music generation systems outperform conventional rule-based and mathematical modelling systems. Among deep learning systems, there are three types of systems that are considered state-of-the-art: transformer systems, VAE-based systems and GANs. The proposed data challenge selects an exemplar system to represent each of these types: MusicTransformer, MusicVAE and MidiNet (GAN). The computer-generated melodies in the development and evaluation datasets are a combination of melodies generated by all three selected systems.

3 Dataset

3.1 Training Data

To investigate whether different music style affects the identification of computer-generated melodies, two datasets are used for training the selected models: Bach Chorales in Music21[1] and pop music from hooktheory[2]. These two training datasets are used for training two separate models for melody generation in this data challenge.

The raw melodies in the datasets are subject to a pre-processing stage. The Bach Chorales dataset contains several voices. Each voice is treated as a separate melody. With regards to pop music in hooktheory, only the melody part is used for training. All melodies are truncated to 32 beats to disregard music structure.

As used by all selected systems [9,15,19], all pre-processed melodies for training are converted into a form of binarised piano roll as demonstrated in Fig. 1. The binarised piano roll represents melodies using a matrix, where each column represents a quarter beat and each row represents a note (such as A4). As each melody has a length of 32 beats and each column represents a quarter beat, the binarised piano roll has 128 ($32 \times 4 = 128$) columns. Moreover, as MIDI files have a pitch number defined between `0` to `127`, there are 128 rows in the binarised piano roll. As a result, the music representation in this paper has a shape of 128×128.

---

[1] https://web.mit.edu/music21/.

[2] https://www.hooktheory.com/.

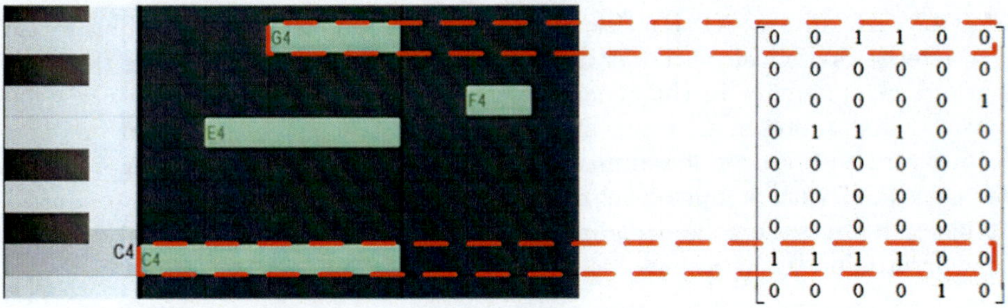

Fig. 1. Binarised Piano roll representation

3.2 Computer-Generated Melodies

As discussed in our brief overview of music generation methods, the selected systems for melody generation are MusicTransformer [9], MusicVAE [15] and MidiNet [19]. In this section, the working principles of these systems are outlined briefly. For more details, the reader is kindly asked to refer to the original papers.

Two datasets (*Music21* and *hooktheory*) are used to train all selected systems twice under the exact configuration hence two models are obtained for each style: Bach and Pop. For each style, one of the resulting models are used to generate melodies in the development dataset, the remaining model is used to generate melodies in the evaluation dataset.

MusicTransformer. MusicTransformer [9] uses a Neural Network Language Model (NNLM) to generate music where the pitch and duration of notes at a time can be considered a word and the motives or phrases can be considered a sentence. This work is among the first using a Transformer to generate music.

Given a sentence S which contains N words w_i, that is, $S = <w_1, w_2, \ldots, w_n> \in V_n$, V_n is the size of the overall vocabulary. The language model aims to find the probability distribution of the sentence, which can be formalised using Eq. (1). Given the forward sequence of a word, the probability of the entire word sequence can be decomposed into the product of the conditional probability of the next word with respect to its forward word. The results of the system show that longer temporal dependencies are well modelled, since repeated or similar phrases can be found in the music generated by the proposed system.

$$P(S) = P(w_1, w_2, \ldots, w_N) = P(w_1)P(w_2|w_1) \cdots P(w_N|w_1, w_2, \ldots, w_{N-1}) \quad (1)$$

The initialisation process of the system depends on the joint probability distribution of the initial sequences hence usually a randomly selected melody with dedicated lengths is used for initialisation. In this data challenge, the effects of the initialisation process for the MusicTransformer are also investigated by examining whether melodies generated by different initialisation seeds can be identified.

MusicVAE. MusicVAE [15] improves the structure of VAE according to the characteristics of music with hierarchical structure, which aims to solve the problem of lack coherence in the generated music when a vanilla VAE is used. The music is first represented using an encoder, constructed with a recurrent neural network to obtain a low-dimensional hidden vector. The resulting vector is then decoded with a multi-level decoder, which reconstructs the vector into a 16-bar unit first, then the decoding process continues with lower-level decoders to generate finer units of melodies.

MidiNet. MidiNet [19] converts music binarised into a piano roll, which is akin to a two-dimensional image. The generator and discriminator of the GAN system then use convolutional neural networks to encode and decode the resulting binarised piano rolls. Besides binarised piano rolls generated by decoders, music composed by humans is also sent to the discriminator for training. At the same time, to maintain coherent connection between the music segments, MidiNet adds information of the front music segment to each layer in the generator. This system is one of the earliest works targeting automatic composition using the method of generating images. It demonstrates the feasibility of using CNN to generate piano roll.

3.3 Human-Composed Melodies

Human composed melodies in this challenge have two sources. Published melodies that are used train the selected music generation systems and unpublished melodies that are required to be composed for this data challenge by university students majoring in music composition.

Published melodies are randomly selected from the dataset that trains the selected music generation systems. The selected melodies are then truncated to 32-beat long segments.

The unpublished melodies in the evaluation dataset is used to test the ability of recognising unknown human-composed music. The data challenge invited professionally trained composers from the China Conservatory of Music to compose a number of melodies. The students were asked to compose melodies in two styles: the Baroque style as composed by J. S. Bach and the common pop style. The structure of the composed melodies is removed with melody truncated to 32-beat long as well.

3.4 Data Representation

The paper uses `pretty_midi`[3] to convert the generated piano roll into a MIDI file which requires the MIDI number and the duration of each note. The MIDI number can be directly indexed by the note. The duration of each beat requires a simple calculation. As each column in the binarised piano roll represents a

[3] http://craffel.github.io/pretty-midi/.

quarter beat, given a tempo value, such as 120 beats per minute (bpm), the duration of each column in the binarised piano roll can be easily calculated.

The instrument selected in the MIDI file is "`Bright_Piano`" with the velocity setting to 127 in MIDI files. The tempi of the MIDI files are randomly selected in the range of 68 bpm, 78 bpm, 88 bpm, 98 bpm, 108 bpm and 118 bpm to avoid the situation where the columns occupied by an individual note would always be the same integer.

3.5 Dataset Formation

Once converted to MIDI files, the computer-generated and human-composed melodies are divided into two datasets: the development dataset and the evaluation dataset. Neither datasets contain labels and they consist of an equal number of Bach-style and pop-style melodies.

In the development dataset, there are 6,000 computer-generated melodies generated by three models. The specific composition of the development dataset is shown in Table 1.

For each type of music generation system, two different datasets were used for training two individual melody generation systems: melodies from Bach Chorales in Music21 (labelled as "Bach" in Table 1) and hooktheory dataset (labelled as "Pop" in Table 1).

Table 1. The development dataset composition of the data challenge where the number in the brackets indicates the number of melodies. "MTrans", "MVAE" and "MNet" represent for music transformer, MusicVAE and MidiNet respectively.

Computer-generated music (6000)					
MTrans (2000)		MVAE (2000)		MNet (2000)	
Bach	Pop	Bach	Pop	Bach	Pop
1000	1000	1000	1000	1000	1000

In the evaluation dataset, there are 4,000 melodies coming from two sources: computer models and human composition.

Among the human-composed melodies, the items truncated from melodies originally used for training music generation systems (labelled as "Training" in Table 2) and specially composed melodies for this data challenge (labelled as "Unpublished") are delineated given the two styles: from Bach Chorales or similar with Bach style (labelled as "Bach" in Table 2) and from hooktheory dataset or common pop style (labelled as "Pop" in Table 2).

The composition of computer-generated melodies are complex. As a general principle, it is necessary to emphasise that the system used to generate melodies in the evaluation dataset and the system used to generate melodies in the development dataset are always different although system architectures may be shared. As the case in the development dataset, each proposed system is trained using two different datasets (labelled as "Bach" and "Pop" in Table 2) hence two separate melody generation systems for different styles are obtained.

Table 2 summarises the composition of the evalution dataset. It is worth mentioning that numbers of melodies generated by MusicTransformer is larger than the other systems in order to investigate the effects of different initialisation configurations. Unlike in the development dataset where only one configuration used for initialisation of the MusicTransformer, the melodies in the evaluation dataset generated by MusicTransformer are the result of three different initialisation configurations, among which one of the initialisation scheme is used in the training process.

Table 2. The evaluation dataset composition of the data challenge where the number in the brackets indicates the number of melodies. "MTrans", "MVAE" and "MNet" represent for music transformer, MusicVAE and MidiNet respectively. The number in the brackets indicates the number of melodies. The title of each column is explained in the context.

Computer-generated melodies (2000)					
MTrans (1200)		MVAE (400)		MNet (400)	
Bach	Pop	Bach	Pop	Bach	Pop
600	600	200	200	200	200
Human-composed melodies (2000)					
Training (1900)		Unpublished (100)			
Bach	Pop	Bach	Pop		
950	950	50	50		

4 Conclusions

The CSMT data challenge requires participants to identify computer-generated melodies among human-composed melodies. The challenge aims to facilitate solutions for determining the source of melodies in possible copyright infringement cases in juridical practice. The term "melody" is used in a limited sense in this data challenge. Melodies were truncated to remove musical structure and they were used without accompaniment. This paper provided an in-depth discussion on the composition and the design of the dataset.

The challenge utilises two components, the development dataset and the evaluation dataset. The development dataset contains only computer-generated melodies whereas the evaluation dataset combines both computer-generated and human-composed melodies. The computer-generated melodies in the development and evaluation datasets are obtained from the same type of systems with slightly different settings. The human-composed melodies were composed specifically for the CSMT data challenge besides existing melodies that were used for system training.

With the presented setup of the challenge, the identification of computer-generated melodies can be considered either an unsupervised outlier detection problem or a supervised classification problem. Both methodologies may suffer from learning the inherent limitations of the selected music generation systems. As a result, the systems proposed by participants in the data challenge may not produce a universally valid approach to identify computer generated melodies, but rely on data distributions instead that characterise state-of-the-art music generation systems. Nevertheless, this approach can still prove to be valuable for practical purposes, as in the legal context introduced earlier, if the models are kept up to date. Moreover, the melody complexity in this data challenge is reduced artificially hence the algorithms from participants may have limited generalisability.

Acknowledgement. The authors acknowledge the contribution from all members in the organisation committee (other than the authors): Prof. ZHANG Ru from Beijing University of Posts and Telecommunications, Dr. LI Zijin from China Conservatory of Music, Mr. ZHU Yidan from Beijing Acoustics Society, Mr. ZHOU Wei from Beijing Zhongwen (Shanghai) Law Firm.

References

1. Ackerman, M., Loker, D.: Algorithmic songwriting with ALYSIA. In: International Conference on Evolutionary and Biologically Inspired Music and Art, Amsterdam, The Netherlands, pp. 1–16. Springer (2017)
2. Briot, J.P., Hadjeres, G., Pachet, F.D.: Deep Learning Techniques for Music Generation. Springer (2020)
3. Brunner, G., Wang, Y., Wattenhofer, R., Zhao, S.: Symbolic music genre transfer with CycleGAN. In: 2018 IEEE 30th International Conference on Tools with Artificial Intelligence (ICTAI), Volos, Greece, pp. 786–793. IEEE (2018)
4. Cheng, T., Smith, J.B., Goto, M.: Music structure boundary detection and labelling by a deconvolution of path-enhanced self-similarity matrix. In: 2018 IEEE International Conference on Acoustics. Speech and Signal Processing (ICASSP), Calgary, Canada, pp. 106–110. IEEE (2018)
5. Dong, H.W., Hsiao, W.Y., Yang, L.C., Yang, Y.H.: MuseGAN: multi-track sequential generative adversarial networks for symbolic music generation and accompaniment. In: Thirty-Second AAAI Conference on Artificial Intelligence, Orlando, USA, pp. 34–41. AAAI (2018)
6. Eck, D., Schmidhuber, J.: A first look at music composition using LSTM recurrent neural networks. Istituto Dalle Molle Di Studi Sull Intelligenza Artificiale **103**, 48 (2002)
7. Goodfellow, I., Pouget-Abadie, J., Mirza, M., Xu, B., Warde-Farley, D., Ozair, S., Courville, A., Bengio, Y.: Generative adversarial nets. In: Advances in Neural Information Processing Systems, Montreal, Canada, pp. 2672–2680 (2014)
8. Herremans, D., Chuan, C.H., Chew, E.: A functional taxonomy of music generation systems. ACM Comput. Surv. (CSUR) **50**(5), 69 (2017)
9. Huang, C.Z.A., Vaswani, A., Uszkoreit, J., Simon, I., Hawthorne, C., Shazeer, N., Dai, A.M., Hoffman, M.D., Dinculescu, M., Eck, D.: Music transformer: generating music with long-term structure. In: International Conference on Learning Representations, Vancouver, BC, Canada (2018)

10. Liu, C.H., Ting, C.K.: Computational intelligence in music composition: a survey. IEEE Trans. Emerg. Top. Comput. Intell. **1**(1), 2–15 (2016)
11. Liu, H.M., Yang, Y.H.: Lead sheet generation and arrangement by conditional generative adversarial network. In: 2018 17th IEEE International Conference on Machine Learning and Applications (ICMLA), Orlando, USA, pp. 722–727. IEEE (2018)
12. Luo, J., Yang, X., Ji, S., Li, J.: MG-VAE: deep Chinese folk songs generation with specific regional styles. In: Proceedings of the 7th Conference on Sound and Music Technology (CSMT), Haerbin, China, pp. 93–106. Springer (2020)
13. Makris, D., Kaliakatsos-Papakostas, M., Karydis, I., Kermanidis, K.L.: Combining LSTM and feed forward neural networks for conditional rhythm composition. In: International Conference on Engineering Applications of Neural Networks, pp. 570–582. Springer (2017)
14. Pinkerton, R.C.: Information theory and melody. Sci. Am. **194**(2), 77–87 (1956)
15. Roberts, A., Engel, J., Raffel, C., Hawthorne, C., Eck, D.: A hierarchical latent vector model for learning long-term structure in music. In: Dy, J., Krause, A. (eds.) Proceedings of Machine Learning Research, Stockholmsmässan, Stockholm, Sweden, 10–15 July 2018, vol. 80, pp. 4364–4373. PMLR (2018)
16. Steedman, M.J.: A generative grammar for Jazz chord sequences. Music Percept. Interdiscip. J. **2**(1), 52–77 (1984)
17. Vaswani, A., Shazeer, N., Parmar, N., Uszkoreit, J., Jones, L., Gomez, A.N., Kaiser, Ł., Polosukhin, I.: Attention is all you need. In: Advances in Neural Information Processing Systems, Long Beach, USA, pp. 5998–6008 (2017)
18. Wang, Y.A., Huang, Y.K., Lin, T.C., Su, S.Y., Chen, Y.N.: Modeling melodic feature dependency with modularized variational auto-encoder. In: 2019 IEEE International Conference on Acoustics Speech and Signal Processing (ICASSP) ICASSP 2019, Brighton, UK, pp. 191–195. IEEE (2019)
19. Yang, L.C., Chou, S.Y., Yang, Y.H.: MidiNet: a convolutional generative adversarial network for symbolic-domain music generation. In: 18th International Society for Music Information Retrieval Conference, Suzhou, China, pp. 324–331 (2017)
20. Zhu, H., Liu, Q., Yuan, N.J., Qin, C., Li, J., Zhang, K., Zhou, G., Wei, F., Xu, Y., Chen, E.: Xiaoice band: a melody and arrangement generation framework for pop music. In: Proceedings of the 24th ACM SIGKDD International Conference on Knowledge Discovery & Data Mining, London, United Kingdom, pp. 2837–2846. ACM (2018)

Research on AI Composition Recognition Based on Music Rules

Yang Deng[1], Ziyao Xu[2(✉)], Li Zhou[3], Huaping Liu[1], and Anqi Huang[1]

[1] NetEase Cloud Music, Hangzhou, China
dynamo@cug.edu.cn, {liuhuaping,huanganqi01}@corp.netease.com
[2] Malong Technologies, Shenzhen, China
ziyxu@malong.com
[3] China University of Geosciences, Wuhan, China
zhouli@cug.edu.cn

Abstract. The development of artificial intelligent composition has resulted in the increasing popularity of machine-generated pieces, with frequent copyright disputes consequently emerging. There is an insufficient amount of research on the judgement of artificial and machine-generated works; the creation of a method to identify and distinguish these works is of particular importance. Starting from the essence of the music, the article constructs a music-rule-identifying algorithm through extracting modes, which will identify the stability of the mode of machine-generated music, to judge whether it is artificial intelligent. The evaluation datasets used are provided by the Conference on Sound and Music Technology (CSMT). Experimental results demonstrate the algorithm to have a successful distinguishing ability between datasets with different source distributions. The algorithm will also provide some technological reference to the benign development of the music copyright and artificial intelligent music.

Keywords: AI composition · Melody arrangement · Machine music creation · Mode recognition

1 Introduction

With the gradual rise of artificial intelligent composition, more and more artificial intelligent composition technology has been introduced for application in the sphere of business. This technology can potentially trigger a series of disputes over copyright issues. For the purpose of managing these potential challenges to intellectual property, it is crucial to design an algorithm that can distinguish between artificial and machine-generated music.

As one of the most important core elements in music, mode plays an important role in judging music [1–3]. Some relevant literature exists that examines the identification algorithm of Chinese modes, but sufficient research on identifying western modes remains to be seen; the context of judgement technology

X. Shao et al. (Eds.): CSMT 2020, LNEE 761, pp. 187–197, 2021.
https://doi.org/10.1007/978-981-16-1649-5_16

for analysing machine-generated artificial music through western modes is a particularly sparse area of research.

In our previous study, a mode-identification algorithm was designed [4], which can classify Chinese traditional modes by constructing a decision-making tree and judging the emotion in Chinese traditional music through identifying modes. The algorithm is consequently shown to have a fairly high accuracy rate for identifying traditional Chinese modes, and thus distinguishing whether or not it is indeed a traditional Chinese mode. While the algorithm's judgement on traditional Chinese modes is fairly accurate, it also exhibits effective anti-interference performance and can successfully identify non-traditional Chinese modes. On this basis, some scholars have constructed a traditional music mode pattern based on traditional Chinese music theory [6], matching the traditional Chinese music modes. The findings indicate that the algorithm has quite a high accuracy rate in identifying traditional Chinese music modes and can distinguish between pentatonic and heptatonic modes.

In previous studies, we have proposed CFCS [5], the chord theory constructor based on the chord construction law and processing logic, and have designed a dynamic programming algorithm for the automatic composition of chords; this enables the realisation of mechanised automatic chord composition. Through experimentation in various cases, the algorithm has been proven to be feasible and effective.

The article proposes OSC (Occidental Scale Constructor) based on a combination of research on traditional Chinese modes and CFCS chord composition function. By constructing the function to conduct mode analysis on monody, the article will make judgements on machine-generated and artificial music based on model stability and abnormal mode changes. Due to the subjectivity and territoriality of music, the range of the study will be limited to popular music based on natural major and minor tunes. The processing of modifier notes such as passing notes, neighbouring nodes, and nonessential notes will not be included.

2 Approach

The main technical issue that the article aims to resolve is the design of an algorithm that can distinguish between artificial and machine-generated music. The adopted technical proposal is to analyse melodic data through a set of western mode construction functions and subsequently make the distinction based on the analytical result.

As shown in Fig. 1, the overall technical pattern of the research can be divided into three parts. Firstly, decode the MIDI byte through a MIDI preprocessing module and divide some characteristic series according to specific music rules. Secondly, analyse the preprocessed files through mode analysis mode to ascertain whether the melody adheres to basic music rules. Finally, identify the data in the last module in accordance with a man-machine identification module, to assess the probability of the melody being either man- or machine-made.

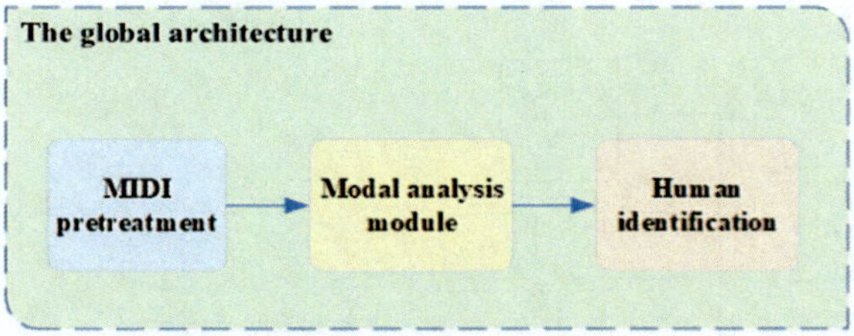

Fig. 1. The overall technical pattern of the proposed approach

MIDI File Preprocessing. MIDI (Musical Instrument Digital Interface) was introduced in the 1980s to amend communication issues between electroacoustic musical instruments, and is currently the most widely accepted music standard format in the composition world; almost all modern music is created and composed using MIDI. As MIDI files usually contain a large amount of information, it is essential to preprocess the MIDI data used in our experiments. Preprocessing mainly involves extracting the scale based on the pitch of the MIDI file, thereby eliminating different interference notes by enumerating the filtration of characteristic intervals and statistical frequency to improve the accuracy of the final result.

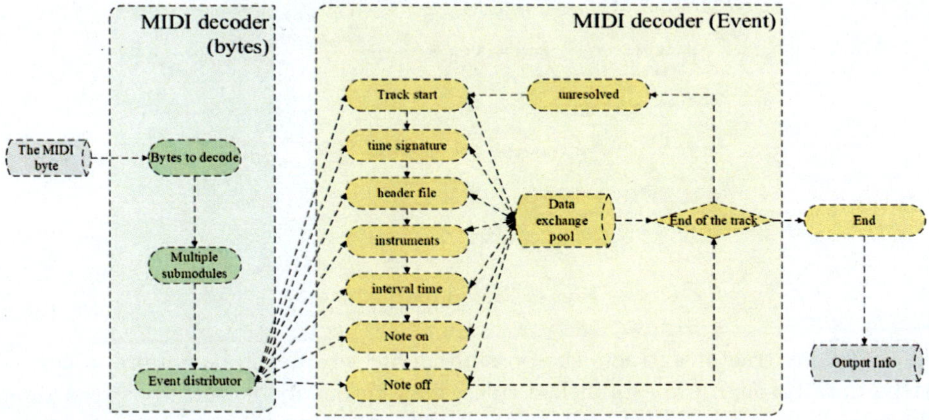

Fig. 2. MIDI file preprocessing (1)

As shown in Fig. 2, the model mainly uses the music's abstract information extracted from the MIDI files for subsequent calculation. It identifies the tracks in the MIDI (accompaniment, drumbeat, melody, polyphony, etc.) based on the established rules before classifying the music construction.

After the MIDI is decoded, the model obtains a series of abstract MIDI information. As demonstrated in Fig. 3, where '0', '1', 'n' denotes the order of

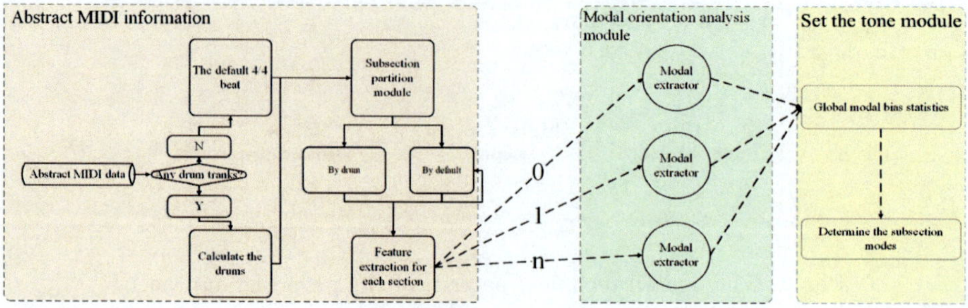

Fig. 3. MIDI file preprocessing (2)

bars, MIDI information is then divided into bars and categorised after data cleansing through a series of classification layers. Finally, each MIDI track goes through modal orientation extraction.

2.1 Modal Extractor

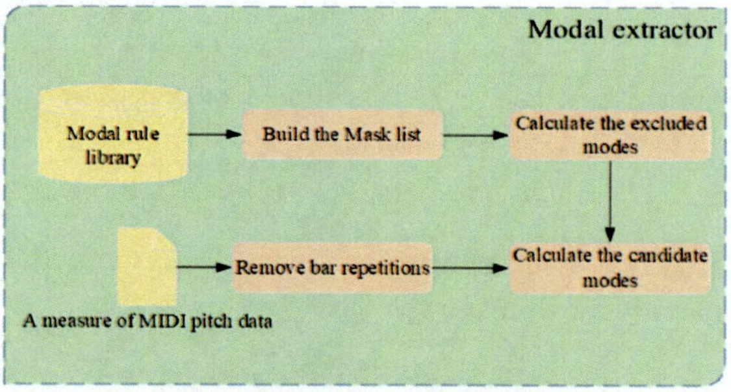

Fig. 4. Modal extractor

The modal extractor extracts the possible mode set of each bar of preprocessed MIDI data through pre-established rules, making bar mode selections regarding the overall most orientated mode. The most frequently used method for extracting the tendentious set is to match the model exclusion mask based on the model rule library generated by OSC and deduce the possible model backward via calculation of the exclusive ones (Fig. 4).

2.2 Modal Rule Library and OSC

Model is a form of organisation structure of music tones with a long-established history of use in practical music. When describing the concept of model, people typically take the pivot note of a model, i.e., the keynote, as the starting and finishing points. Other notes will be arranged in the form of a scale, based on the sequence of the pitch. This is known as modes.

The natural major and minor are the most common modes in the western modal system and in pop music to this day. The article proposes Occidental Scale Constructor (OSC) and constructs the model rule library based on the composition system of natural major and minor modes.

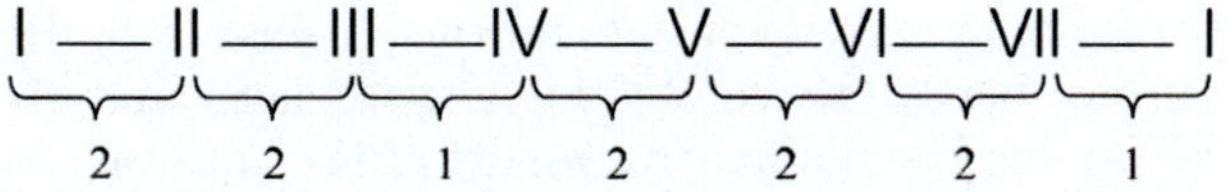

Fig. 5. The composition principle of natural major

The Constructor of Natural Major. The natural major is a scale system consisting of two whole tones, a semitone, three whole tones, and a semitone. See Fig. 5, where '2' denotes a whole tone and '1' denotes a semitone. Starting from any note, any scale system that is constructed in accordance with the aforementioned rules can be called a natural major system.

Based on the rules above, the construction function of the natural major can be formulated as:

$$\begin{aligned} F_{Major}(S,O) = [&S + (O * 12), S + 2 + (O * 12), S + 4 + (O * 12), \\ &S + 5 + (O * 12), S + 7 + (O * 12), S + 9 + (O * 12), S + 11 + (O * 12)]. \end{aligned} \tag{1}$$

Under the mapping relation F (function), S (step) in the function refers to any given sound level, while O (octave) represents the octave group. The natural major scale of current group can thus be constructed.

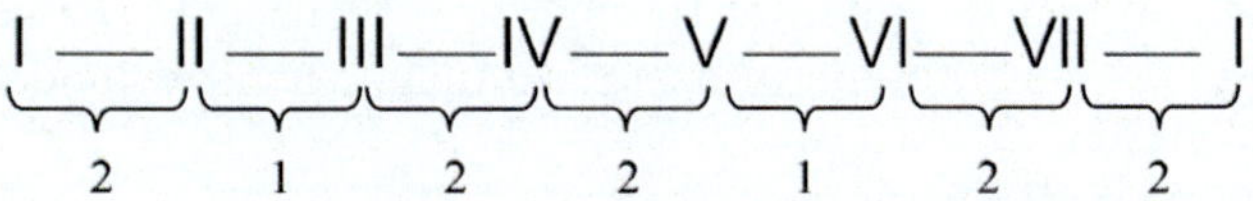

Fig. 6. The composition principle of natural minor

The Constructor of Natural Minor. The constitution system of the natural minor is a whole tone, a semitone, two whole tones, a semitone, and two whole tones. See Fig. 6.

According to Eq. (1), the construction function of the natural minor key will therefore be:

$$\begin{aligned} F_{Minor}(S,O) = [&S + (O * 12), S + 2 + (O * 12), S + 3 + (O * 12), \\ &S + 5 + (O * 12), S + 7 + (O * 12), S + 8 + (O * 12), S + 10 + (O * 12)]. \end{aligned} \tag{2}$$

2.3 Mask Remove Algorithm

It is extremely unlikely that the melody of a single bar would exhibit the complete scale of all models. For example, when there is any black key note, only C natural major can be excluded while all remaining models can still potentially become the dominative model of the entire piece. Based on this issue, it is possible to construct an excluding M (masking) for the melody of a given bar based on the constitution system of natural major and minor. Conducting model-exclusive calculations on the pitch is also an option, for the purpose of obtaining all variant models of the current bar before conducting a systematic analysis on all variant models and ascertaining the dominative model of the entire piece.

The mask sequence based on the major will be constructed as such:

$$\begin{aligned} M_{major}(S,O) = [&S+1+(O*12), S+3+(O*12), \\ &S+6+(O*12), S+8+(O*12), S+10+(O*12)]. \end{aligned} \tag{3}$$

Compared with the natural major, the scale of natural minor elevates the fifth scale on the foundation of the natural major. Consequently, the mask sequence construction function of minor will be:

$$\begin{aligned} M_{minor}(S,O) = [&S+1+(O*12), S+3+(O*12), \\ &S+6+(O*12), S+7+(O*12), S+10+(O*12)]. \end{aligned} \tag{4}$$

If the scale in MMinor (S,O) is not evident in some bars, the affiliated minor of the major whose key note is S can be adopted as the alternative model of the current bar.

Under the mapping relation of the M (Mask), with given S (Step) and O (Octave), the exclusive sequence of the natural major that uses S as keynote can be obtained. When the pitch of the bar is in this sequence, we can exclude this model. Taking C natural major as an example, when the model is C natural major and S = 0, then:

$$M_{major} = [1+(O*12), 3+(O*12), 6+(O*12), 8+(O*12), 10+(O*12)]. \tag{5}$$

If in some bars, Pitch-13, it can calculate the scale of O based on the twelve-tone equal temperament. And when O = 1, then:

$$M_{major}(0,1) = [13, 15, 18, 20, 22] \tag{6}$$

Therefore,

$$Pitch \in M_{major}(0,1). \tag{7}$$

According to the above results, it can be concluded that the current bar does not belong to C natural major. After excluding all impossible models based on each bar of the piece in its entirety, the set of all possible models of the current bar can be obtained. After statistically analysing all alternative models, the model's tendency sequence can then be calculated. Based on the model tendency, it would be possible to filtrate the alternative models of all bars.

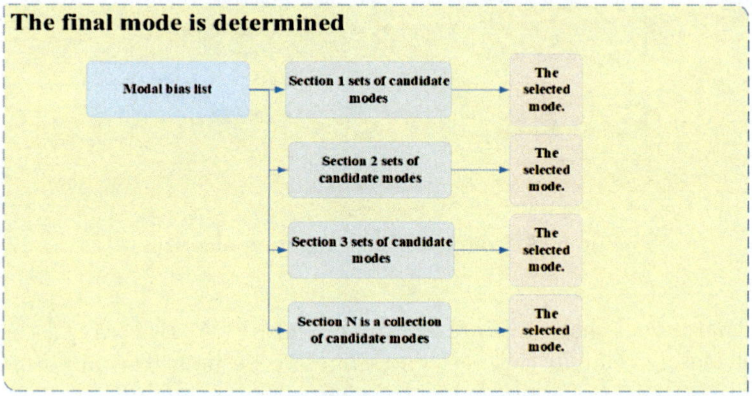

Fig. 7. Mode determination

For example, through calculation, it is possible to determine that the model tendency sequence list of a piece is [C major, G major, D major...] and the alternative model set of the first bar is [G major, A major, E major...]. Consequently, if one were to make the choice based on the sequence in the list, the result would be G major. Likewise, by selecting the model for all bars, the model tendency of the whole piece would thus be obtained (Fig. 7).

2.4 AI Composition Recognition

One of the most significant features of music is model stability. Although many musicians commit themselves to breaking the regular model system and discovering new creation techniques, mainstream music currently still adopts the stable model. Even the modulation or detune obeys certain rules and frequency. For example, modulation usually occurs between closely related models, as frequent or distant modulation would influence the stability of the music. Therefore, the article designs an algorithm to judge abnormal models and consequently attain the statistics of the abnormal model change, so as to judge the probability of the music being artificial or machine-made.

Figure 8 illustrates the technological flow chart that can be adopted to judge man-made or machine-made property through abnormal model change. This abnormal model change usually takes the form of unconventional modulation or with uncertain model. For instance, the models of the bars in one melody are identified as [C, C, C, G, G, E, A, B, F, A, C]. The former five bars are [C, C, C, G, G]. The transmission from C to G belongs to close modulation, so there is

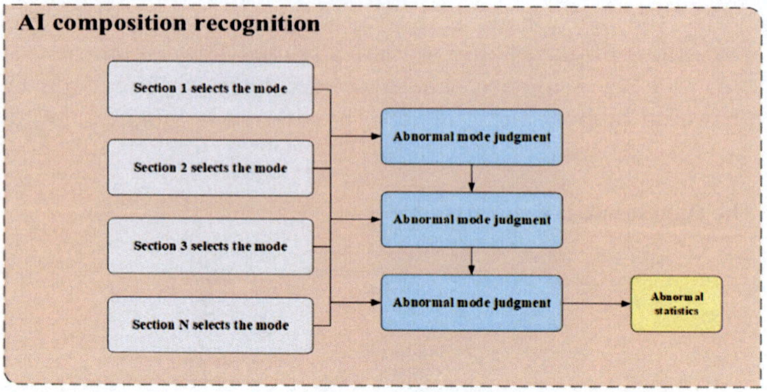

Fig. 8. AI composition recognition algorithm

no abnormal model change. However, the models [E, A, B, F, A, C] that follow it are not closely related; this case can therefore be judged as abnormal model change. Six instances of abnormal model change can be identified in this melody, while there are ten instances when the model can be modified. Thus, the output score of the melody is $6/10 = 0.6$.

3 Experiment

The data used are provided by CSMT [7]. The development dataset contains 6000 MIDI files with monophonic melodies generated by artificial intelligence algorithms. The tempo is between the 68bpm and 118 bpm (beat per minute). The length of each melody is 8 bars, and the melody does not necessarily include complete phrase structures. The evaluation dataset contains 4000 MIDI files with exact configurations of development dataset with two exceptions: 1) A number of melodies composed by human composers are added, 2) There are a number of melodies generated by algorithms with minor difference compared to the algorithms in the development dataset.

Experimental results on CSMT datasets indicate that the score distributtion of the development data is obviously at a low level(Fig. 9), while the score distributtion of the evaluation data is obviously at a high level (Fig. 10).

We summarize the Area Under Curve (AUC) scores for AI composition recognition on the CSMT evaluation dataset in Table 1. A general observation we can draw from the results is that our proposed algorithm has achieved good performances and stability across different styles, generation systems and publish statuses. Significantly, we reach 0.9868 AUC on the melodies generated by GAN. The overall AUC also proves the effectiveness of our method.

Through experiments on 10,000 samples, our algorithm shows a successful identification performance on the judgment of man- or machine-made works. However, complex composing techniques and the evaluation of the time value of notes are not be included. Under the circumstance of short duration time, the

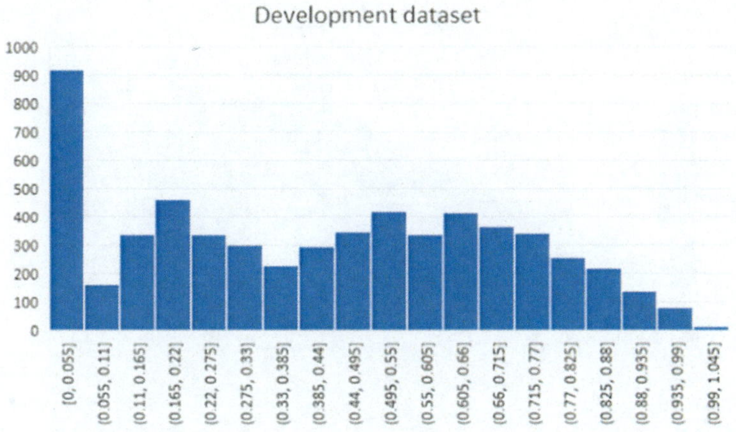

Fig. 9. Score distributtion of the development data. The x axis denotes the score, whereas the y axis denotes the number of examples in the development dataset.

Table 1. Area Under Curve (AUC) scores for AI composition recognition

Taxonomy		AUC
Styles	Bach	0.7731
	Pop	0.7614
Generation systems	GAN [8]	0.9868
	Transformer [9] (length of initial sequences: 16)	0.7620
	Transformer [9] (length of initial sequences: 32)	0.6828
	Transformer [9] (length of initial sequences: 64)	0.7346
	Transformer [9] (overall)	0.7265
	VAE [10]	0.6468
Publish statuses	Published	0.7621
	Unpublished	0.7709
Overall		0.7626

melody created by human and machine can not be clearly judged by composition techniques and abstract rules such as musical form structure. A small number of melody pieces can not be clearly judged even by professionals. However, considering that the melody itself has a certain flexibility, there is no strict unified standard, so the experimental results prove that the algorithm is effective and feasible.

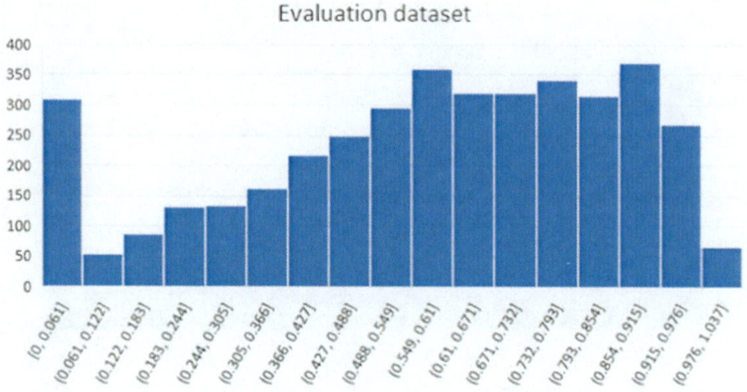

Fig. 10. Score distributtion of the evaluation data. The x axis denotes the score, whereas the y axis denotes the number of examples in the evaluation dataset.

4 Conclusion

Starting from the music mode recognition and the essence of the music, the article proposes Occidental Scale Constructor based on the CFCS chord constructor. The article also constructs a mode-based music-rule-identifying algorithm through combining OSC with the mask remove algorithm, which will identify the mode stability and abnormal mode change, to judge whether the piece is machine-generated. Experimental results on CSMT datasets demonstrate the algorithm to have a successful identification ability of machine-generated music. The algorithm will also provide some technological reference to the benign development of the music copyright and artificial intelligent music.

References

1. Faraldo, Á., Gómez, E., Jordà, S., Herrera, P.: Key estimation in electronic dance music. In: Ferro, N., et al. (eds.) Advances in Information Retrieval. ECIR 2016. LNCS, vol. 9626. Springer, Cham (2016). https://doi.org/10.1007/978-3-319-30671-1_25
2. Korzeniowski, F., Widmer, G.: End-to-end musical key estimation using a convolutional neural network. In: the 25th European Signal Processing Conference (EUSIPCO-2017) (2017)
3. Korzeniowski, F., Widmer, G.: Genre-agnostic key classification with convolutional neural networks. In: The 19th International Society for Music Information Retrieval Conference (2018)
4. Deng, Y., Zhou, L., Ni, S., Zhang, S., You, M.: Research on the recognition algorithm of Chinese traditional scales based on decision tree. In: The 37th Chinese Control Conference (2018)
5. Deng, Y., Zhou, L., Xu, D., Yue, C., You, M., Zhou, R.: Study on adaptive chord allocation algorithm based on dynamic programming. J. Fudan Univ. (Nat. Sci.) **58**, 393–400 (2019)

6. You, M., Chen, L., Zhou, L., He, J.: Research on modal identification of Chinese folk music based on template matching. J. Fudan Univ. (Nat. Sci.) **59**, 262–269 (2020)
7. Li, S., Jing, Y., Fazekas, G.: A novel dataset for the identification of computer generated melodies in the CSMT challenge (2020)
8. Yang, L.-C., Chou, S.-Y., Yang, Y.-H.: MidiNet: a convolutional generative adversarial network for symbolic-domain music generation. In: The 18th International Society for Music Information Retrieval Conference (2017)
9. Anna Huang, C.-Z., Vaswani, A., Uszkoreit, J., Simon, I.: Music transformer: generating music with long-term structure. In: The International Conference on Learning Representations (2018)
10. Roberts, A., Engel, J.H., Raffel, C., Hawthorne, C., Eck, D.: A hierarchical latent vector model for learning long-term structure in music. In: The Proceedings of Machine Learning Research (2018)

A Transformer Based Pitch Sequence Autoencoder with MIDI Augmentation

Mingshuo Ding and Yinghao Ma(✉)

Peking University, Beijing 100871, China
{dingmingshuo,yhma625}@pku.edu.cn

Abstract. Despite recent achievements of deep learning automatic music generation algorithms, few approaches have been proposed to evaluate whether a single-track music excerpt is composed by automatons or Homo sapiens. To tackle this problem, we apply a masked language model based on ALBERT for composers classification. The aim is to obtain a model that can suggest the probability a MIDI clip might be composed condition on the auto-generation hypothesis, and which is trained with only AI-composed single-track MIDI. In this paper, the amount of parameters is reduced, two methods on data augmentation are proposed as well as a refined loss function to prevent overfitting. The experiment results show our model ranks 3rd in all the 7 teams in the data challenge in CSMT (2020). Furthermore, this inspiring method could be spread to other music information retrieval tasks that are based on a small dataset.

Keywords: ALBERT · Autoencoder · MIDI truncation · Small dataset

1 Introduction

Methods based on machine learning have been widely proposed for automatic music generation since significant progress on deep learning. Nowadays, more and more melodies can be composed by artificial intelligence through using the pitch and length of the notes in human music as primary inputs to mimic humans [1–3]. Unlike checking counterpoint in multi-track melodies and evaluation self-similarity matrix in music structure analysis, few objective algorithms or indicators have been put forward to assess whether a single-track short melody is created by a machine or a person. Although several attempts has been made, such as measures from information theory to compare Bach's music [4], or probability transfer relation with the N-gram model to compare British and American folk music melody [3], most of the classification model on composers works are based on human opinions, namely, the participants listened to a music excerpt and then judged whether it was composed by a human or an AI [5–8].

However, the result of listening tests might contain individual or group differences, which makes them difficult to be compared among different people, especially when the amount of samples is small. Finding a relatively common

X. Shao et al. (Eds.): CSMT 2020, LNEE 761, pp. 198–207, 2021.
https://doi.org/10.1007/978-981-16-1649-5_17

and objective approach to classify the composer of a short piece of melodies in various musical styles can make different music generation models comparable. The purpose of this study is to find an objective and effective method to generate an indicating value of whether a music clip is human-composed by analyzing the AI-made melodies.

Features extracting is an essential component for music series related tasks. For the single-track data without chords, there are some methods that rely on N-gram [3,9]. However, this approach is difficult to model the long-term dependence and the following dependence, and the data is sparse with the exponential growth of probability as sequence length increases, which leads to poor generalization ability. Besides, Bidirectional Encoder Representations from Transformers (BERT, Fig. 1) [10] might be a promising technique except its large amount of parameters such as learning an embedding for a sequence after parameters reduction [3].

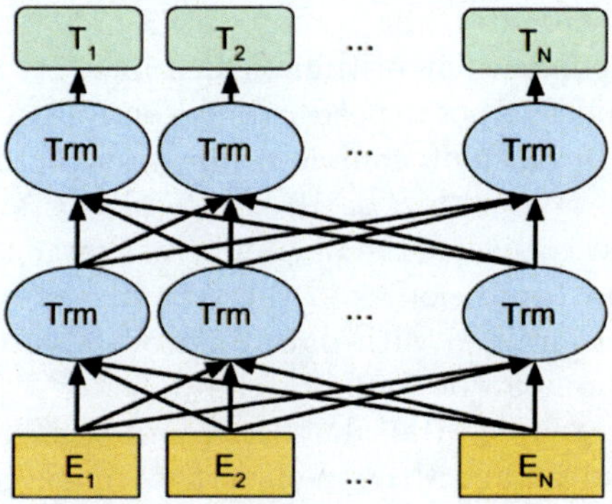

Fig. 1. BERT uses a bidirectional transformer. [10]

In fact, BERT as a pre-trained models [10–12] has dominated the field of Natural Language Processing (NLP) in the past two years. This model uses self-supervised learning to encode contextual information to obtain a powerful and universal representation. This representation can improve performance, especially in situations where data for downstream tasks is limited. More recently, BERT-like models have been applied to speech processing [13–16]. However, such models usually maintain a large number of parameters in both speech tasks and text tasks, requiring a large amount of data and memory for training and computation. Therefore, it might be prone to overfit when pre-training data is relatively scarce, such as in music related cases.

A Lite BERT (ALBERT) [17] is a simplified version of BERT that shares the same parameters at all layers and decompose the embedding matrix to reduce most of the parameters. Although the number of parameters is reduced, the representation learned in ALBERT is still robust and task agnostic, so that ALBERT can achieve similar performance to BERT in the same downstream task [18], thus is also regarded as obtaining characteristics about the input itself.

In this paper, a masked language model (MLM) which is based on ALBERT is introduced into MIDI processing and a new self-supervised model is proposed.

The rest of this article is organized as follows. In Sect. 2, the dataset used in the study is described as well as the data preprocessing and strategies used for data augmentation. In Sect. 3, the pipeline of the research, the methods on prevention of overfitting are demonstrated, as well as the detail of the ALBERT model and the approach to evaluate the probability of each composer. Section 4 covers the main experimental processes and results. The fifth section we have made the summary and the prospect.

2 Dataset

2.1 Training Data

The data set is provided in the data challenge of Conference of Sound and Music Technology (CSMT) 2020 [19]. The training data only contains the music generated by artificial intelligence algorithms which includes 6000 MIDI files. Each file is single melodic music whose speed is between 68BPM and 118BPM. Each melody is 8-bar length, without complete phrase structure. In fact, complete music sentences are always with 8 or 16 bars and this suggests that the start point of each music excerpt is not the beginning of any music sentences. Besides, it should be noted that the melodies in the training data set are generated by several machine models trained with data in two unannounced different music genres. More information can be found at the website[1].

Despite many open source MIDI datasets on the internet such as the one on reddit with 3.65 GB multi-track MIDI in all sorts of music genre[2], the single-track music clips like what is provided in the data challenge are rare, not to mention the uncertainty on music genre. As a consequence, it is difficult to extract a convincing main melody especially condition on similar music range and notes distribution. Therefore, training did NOT use any human composed data.

2.2 Data Preprocessing

For the specific problem of comparing the similarities of melodies, the rhythm and pitch are important characteristics, since people usually pay attention to them when they perceive music melodies [20]. Thus, the MIDI sequence of 8 bars can be segmented into 128 hexadecimal notes or 256 thirty-second notes, as the speed and the starting and ending time of the notes are marked. Whether the unit of the 8-bar music is a hexadecimal note or a thirty-second note depends on the shortest note length in the given MIDI, and there are 256 notes or so in a music sequence for most of the cases. Considering the fact that it is meaningless in music to divide a quarter note into twelve equal parts in the vast majority

1 http://www.csmcw-csmt.cn/data/2020/ai-composition-recognition2020/?from=timeline.

2 https://www.reddit.com/r/datasets/comments/3akhxy.

of cases, there is no musical necessity to do so except for compatibility with the relative rarity of triplets and sixteenth notes. Thus, we classify all triplets as three quavers or three sixteenth notes in the same probability, which leads to the total length of a music sequence not being 256. Given that the speed of each music piece is uniformed as the tempo of each music piece is similar to Andrate, the feature of speed in each MIDI sequence is not taken into consideration. In this way, each single-track MIDI clip is turned into a pitch sequence.

2.3 Data Augmentation

Although a noticeable amount of parameters has been decreased in ALBERT relative to BERT parameters, 6000 MIDI data are somehow relatively poor for training. As a consequence, it is vital to adopt some measures on data augmentation. Unfortunately, data augmentation methods usually used in NLP tasks [21] can be seldom used in music series processing.

Randomly swapping is a common approach, but the exchange of music notes may cause non-negligible differences in feeling for a human listener. Music clips for the composition of humanity, for example several sixteenth notes in a crotchet or half note exchange with other sounds, could lead to a strange auditory experience, and let the audience regard the music piece as machine-created. Synonym replacement is not suitable in a sequence of music analysis, because there is no specific semantic like natural language for music notes or sequences. Therefore, it's hard to define whether two notes are "synonym". Even replacing the octave "synonym" is unacceptable in a lyrical semiquaver with a long note, which results in a clear change in music expressed in human emotion, though little differences infrequency spectrum. In addition, Random insert and delete run a high risk which could make melody strange and weird. It is also hard to change the music from major mode to minor mode for augmentation because the mode is hard to find with only single-track especially without music sentences in it. Moreover, the tempo change augmentation can be hardly used either as the tempo is already uniformed. So we proposed two methods to augment data.

Transposition. The first data augmentation measure taken in our research is transposition in music tunes. Since music does not make a significant difference, at least not in the respect whether it is generated by human beings or artificial intelligence if it is just changed in music mode.

Each time, a transposition raises or lowers all the notes in the same pitch sequence by a same random music interval. All the positions the MIDI clips might be transposed to is restricted by both the MIDI range 128 and the music range, that is the highest note subtract the lowest note. The number of cases for a certain music piece *num* is as follows, including zero transposition:

$$num = 128 - highest + lowest + 1. \quad (1)$$

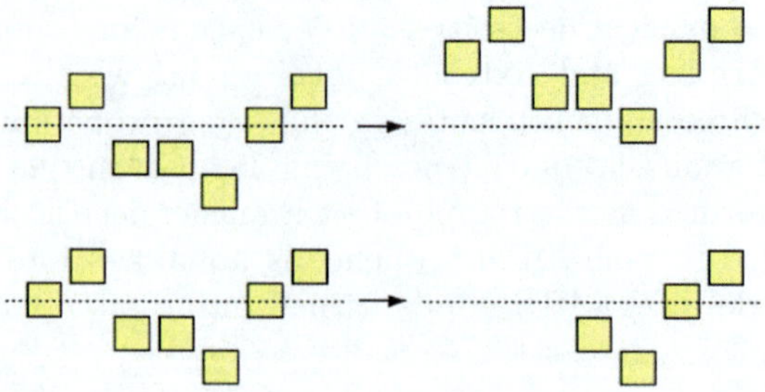

Fig. 2. Data augmentation approaches: transposition and random truncation

Each MIDI transposition is implemented with the same possibility to all the cases. In this way, several relatively same melodies in different music tunes are generated by the transposition data augmentation.

Random Truncation. In addition, BERT's training results contain position embedding and thus absolute position information [22], for example the word at the beginning of the sentence may be regarded as the subject of the sentence. But the dataset neither includes complete phrase information nor cadence in multi-track, therefore, some location information in the training set retained by BERT belongs to some kind of over-fitting. In order to give up this information, we randomly delete the first few notes of each pitch sequence for the model.

3 Methods

The pipeline of our model is shown in Fig. 2. First of all, the training set will undergo a data preprocessing part as described above and be expanded by the two data augmentation approaches. Secondly, a MLM task based on ALBERT is trained with refined loss function for an autoencoder on the expanded training set. Lastly, the trained model will be used for evaluation.

3.1 Avoid Over-fitting

Since there is only machine-generated data used and no data on human composition, it is still easy to overfit even after data augmentation. To cope with this problem, several additional measures have been taken to prevent from data overfitting.

Refined Loss Function. Some studies have shown that slight adjustment of the loss function l can prevent overfitting greatly [23]:

$$l_{new} = |l_{origin} - b| + b, \tag{2}$$

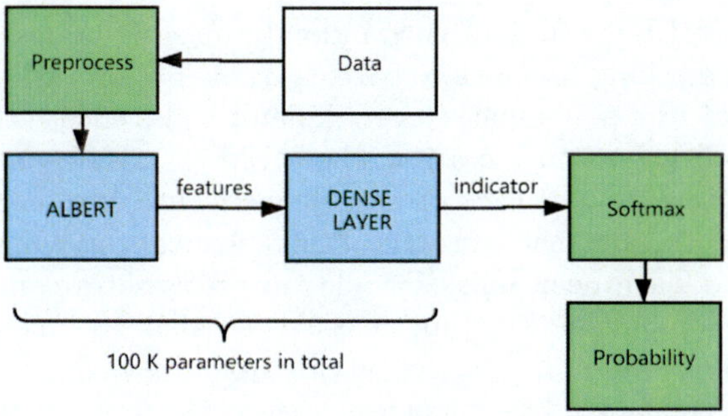

Fig. 3. Flow chart of data processing

where b is a little positive real parameter which is problem related. The model is trained with the refined loss function and b is set to 0.05 which is a magic number in some NLP tasks to prevent from pursuing zero-value of original loss function but only to a close-zero value.

Smaller Transformer. The number of parameters in the BERT model is extremely large. Even in the ALBERT model using shared parameters, the number of parameters can easily lead to overfitting on such a small dataset. Therefore, on the basis of retaining the structure of ALBERT, the dimension of embedding is 64, the number of multi-layers is set to 2 as well as the number of multi-head is 4. As a result, the amount of parameters of ALBERT is reduced significantly to around 103.6k, thus avoiding potentially overfitting on the training set.

3.2 Training Method

There are two important tasks of Bert's training process [10]: Masked Language Model (MLM) and Next Sentence Prediction (NSP). However, the NSP task is not necessary in this problem, because the training set does not include complete phrase information. Actually, it will be hard to divide notes into several phrases. On the contrary, MLM is suitable to tackle this problem.

We hope that the AI composing algorithms used in the dataset which is relatively certain can be fitted through the coding representation obtained by the more "universal" ALBERT with a large number of parameters. Some items of the MIDI sequence is masked and predictions are made on each of the masked note position based on the corresponding embedding vector learned by ALBERT. Such predictions might be closer to the results of some of the AI composers than to those of humans. Assuming that the music was composed by an algorithm fitted by the ALBERT model, the average "probability" of each masked note being the same as its ground truth note can be seen as the "P-value" indicating whether it was created by AI and the hypothesis shall be accept or reject.

Note that ALBERT training will randomly mask N-grams to make predictions [17]. If the masking happens to cover a whole bar or a whole chord formed by adjacent notes, the notes masked are difficult to be effectively predicted.

After comprehensive consideration, the MLM task is the only used task for training. Each time, about 15% of the elements has been randomly masked in a pitch sequence, and then use the other elements not masked to predict the elements that have been masked. Selecting 15% notes can ensure that the essential music components are not masked, so that the model can produce effective prediction, and random selection can avoid overfitting to a certain extent as well. And the softmax cross-entropy is used as the loss function of the model to evaluate the distance between the one-hot vector ground truth and the 128 dimensions vector representing the probability of being each of the 128 MIDI notes, followed by the process mentioned above to refine the loss.

3.3 Evaluation

When evaluating, for a pitch sequence, each note will be masked successively. Then, the probability p_i of the ith masked note is predicted by the trained ALBERT, and the average probability of all notes is the probability that this data is composed by AI. Formally, the number of notes in this pitch sequence is denoted as n, and suggests the probability of AI generating is as follows:

$$p = \frac{1}{n}\sum_{i=1}^{n} p_i \tag{3}$$

Thus, the probability of each data created by humans, which this task required, can be obtained by $1 - p$.

4 Experiment

4.1 Data Setup

Based on the Albert model, the autoencoder model is trained with MLM tasks on the dataset provided by CSMT (2020) after augmenting. Both data augmentation strategies mentioned above are used for all the data in the training set.

Firstly, we use *pretty_midi* [24] reads the data in and then preprocesses it. For a pitch sequence after preprocessing, 31 different transpositions are generated including the case remaining the same. And16 of them are implemented with different values of random truncation range in 1to 100. Due to the fact that there are only 12 different modes in an octave and the limitation of computing resources, the size of the augmentation is not extremely large and only part of them are used for training. Therefore, the size of the training set is expanded to 186000, which is enough for training on the small ALBERT.

4.2 Environment and Hyper Parameters

Under the good parameter control strategy, the Albert is able to be deployed on a GTX 1050Ti NVIDIA graphic card. *Pytorch* [25] and *Hugging Face* [26] are used in the process of building and training the algorithm. The small batch size is 64 and the default learning rate is 10^{-3} with AdamW optimizer [27]. The parameter b mentioned is set as 0.05. Because there is no ground truth in the test set, we can not carry out the ablation experiment, the selection of hyper parameters is all based on past experience.

4.3 Experiment Result

The data challenge uses the average under receiver operating characteristic curve (AUC) as an indicator for each model performance. The overall performance of AUC is 0.6821 which is rank 4th in the 9 models including the baseline model and rank 3rd in all of the 7 teams that finished the data challenge.

The details of the result are shown in the following table (Table 1, Table 2 and Table 3).

Table 1. The AUC of test data in different music style

Style	AUC value
J.S. Bach	0.6984
Pop song	0.6673

Table 2. The AUC of test data composed by different AI algorithm

Algorithms	AUC value
GAN	0.7458
Transformer	0.7811
VAE	0.3210

Table 3. The AUC of test data composed by human

Category	AUC value
Published	0.6895
Unpublished	0.5404

The AUC values of different music styles do not show significant difference, which implies our model may keep an objective evaluation among different music styles. Furthermore, the result of VAE composed is extremely low, even worse than the random guess. Although the test data is not published and audios can not be listened for finding some missing patterns, this phenomenon deserves more attention. Finally, the unpublished result is a bit lower than the published data. This might be caused by the relatively small number of unpublished data and these data are composed by conservatory students instead of composers like Bach and these might keep some difference with each other.

5 Conclusion

In this paper, we proposed an autoencoder approach based on ALBERT with the aim to set up an indicator to reject the hypothesis that the music excerpt is composed by machine. The ALBERT model is trained self-supervised with a MLM to mimic the AI-composer. Experimental results confirmed that the brand-new method outperforms some of other algorithms and rank 3rd and shows little difference in two music styles. Besides, we found the model performance on VAE models is extremely low, therefore, deserve more attention.

Our model provides a meaningful approach and can be spread to similar tasks with small dataset. However, there are several problems unavoidable as well. To begin with, the whole semantics of the encoder is hard to be understood as the performance on some of the models is relatively high and others are extremely low, which suggest the obvious uncertainty on there liability of the workflow. In addition, the indicator in our model based on the encoder works in the way of p-value and keeps some weakness by nature. Some good music pieces may have high probability to be composed by both human composers and artificial intelligence and other weird MIDI clips might be low possibility to be composed by both homo sapiens and automatons. These unsolid pseudo p-values shall be avoided or be implemented in great caution when it is spread to other tasks if there are some data in another class.

References

1. Liu, C.H., Ting, C.K.: Computational intelligence in music composition: a survey. IEEE Trans. Emerg. Top. Comput. Intell. **1**(1), 2 (2016)
2. Dong, H.W., Hsiao, W.Y., Yang, L.C., Yang, Y.H.: arXiv preprint arXiv:1709.06298 (2017)
3. Li, Z., Li, S.: Proceedings of the 7th Conference on Sound and Music Technology (CSMT), pp. 121–130. Springer (2020)
4. Ren, I.Y.: ECE Department, University of Rochester (2015)
5. Liang, F.T., Gotham, M., Johnson, M., Shotton, J.: ISMIR, pp. 449–456 (2017)
6. Chu, H., Urtasun, R., Fidler, S.: arXiv preprint arXiv:1611.03477 (2016)
7. Huang, A., Wu, R.: arXiv preprint arXiv:1606.04930 (2016)
8. Unehara, M., Onisawa, T.: 10th IEEE International Conference on Fuzzy Systems.(Cat. No. 01CH37297), vol. 3, pp. 1203–1206. IEEE (2001)
9. Ogihara, M., Li, T.: ISMIR, pp. 671–676 (2008)
10. Devlin, J., Chang, M.W., Lee, K., Toutanova, K.: arXiv preprint arXiv:1810.04805 (2018)
11. Peters, M.E., Neumann, M., Iyyer, M., Gardner, M., Clark, C., Lee, K., Zettlemoyer, L.: arXiv preprint arXiv:1802.05365 (2018)

12. Radford, A., Narasimhan, K., Salimans, T., Sutskever, I.: Improving language understanding by generative pre-training (2018)
13. Liu, A.T., Yang, S., Chi, P.H., Hsu, P.C., Lee, H.: ICASSP 2020-2020 IEEE International Conference on Acoustics, Speech and Signal Processing (ICASSP), pp. 6419–6423. IEEE (2020)
14. Jiang, D., Lei, X., Li, W., Luo, N., Hu, Y., Zou, W., Li, X.: arXiv preprint arXiv:1910.09932 (2019)
15. Ling, S., Liu, Y., Salazar, J., Kirchhoff, K.: ICASSP 2020-2020 IEEE International Conference on Acoustics, Speech and Signal Processing (ICASSP), pp. 6429–6433. IEEE (2020)
16. Schneider, S., Baevski, A., Collobert, R., Auli, M.: arXiv preprint arXiv:1904.05862 (2019)
17. Lan, Z., Chen, M., Goodman, S., Gimpel, K., Sharma, P., Soricut, R.: arXiv preprint arXiv:1909.11942 (2019)
18. Chi, P.H., Chung, P.H., Wu, T.H., Hsieh, C.C., Li, S.W., Lee, H.: arXiv preprint arXiv:2005.08575 (2020)
19. Li, S., Jing, Y., Fazekas, G.: arXiv preprint arXiv:2012.03646 (2020)
20. Kim, Y.E., Chai, W., Garcia, R., Vercoe, B.: ISMIR (2000)
21. Wei, J., Zou, K.: arXiv preprint arXiv:1901.11196 (2019)
22. Lee, J., Lee, Y., Kim, J., Kosiorek, A., Choi, S., Teh, Y.W.: International Conference on Machine Learning (PMLR, 2019), pp. 3744–3753 (2019)
23. Ishida, T., Yamane, I., Sakai, T., Niu, G., Sugiyama, M.: arXiv preprint arXiv:2002.08709 (2020)
24. Raffel, C., Ellis, D.P.:15th International Society for Music Information Retrieval Conference Late Breaking and Demo Papers, pp. 84–93 (2014)
25. Paszke, A., Gross, S., Massa, F., Lerer, A., Bradbury, J., Chanan, G., Killeen, T., Lin, Z., Gimelshein, N., Antiga, L., et al.: Advances in Neural Information Processing Systems, pp. 8026–8037 (2019)
26. Wolf, T., Debut, L., Sanh, V., Chaumond, J., Delangue, C., Moi, A., Cistac, P., Rault, T., Louf, R., Funtowicz, M., Davison, J., Shleifer, S., von Platen, P., Ma, C., Jernite, Y., Plu, J., Xu, C., Scao, T.L., Gugger, S., Drame, M., Lhoest, Q., Rush, A.M.: arXiv preprint arXiv:1910.03771 (2019)
27. Loshchilov, I., Hutter, F.: arXiv preprint arXiv:1711.05101 (2017)

Author Index

X. Shao et al. (Eds.): CSMT 2020, LNEE 761, pp. 209–210, 2021.
https://doi.org/10.1007/978-981-16-1649-5